MODERN MILK PRODUCTION

MODERN MILK PRODUCTION

*Its Principles and Applications for
Students and Farmers*

MALCOLM E. CASTLE and
PAUL WATKINS

*With a Foreword by
Professor William Holmes*

faber and faber
LONDON·BOSTON

First published in 1979
by Faber and Faber Limited
3 Queen Square London WC1
Second edition 1984
Set by Latimer Trend & Company Ltd Plymouth
Printed in Great Britain by
Redwood Burn Ltd, Trowbridge
All rights reserved

British Library Cataloguing in Publication Data

Castle, Malcolm E.
  Modern milk production. — 2nd ed.
  1. Dairying
  I. Title   II. Watkins, Paul, 1921–
  637'.1   SF239
  ISBN 0-571-13242-1

Library of Congress Cataloging in Publication Data

Castle, Malcolm E.
  Modern milk production.

  Includes index.
  1. Dairy farming.   2. Dairying.   I. Watkins, Paul.   II. Title.
  SF239.C34   1984   637'.1   83-20656
  ISBN 0-571-13242-1 (pbk.)

CONTENTS

Frequency of Milking—Milking Machines—The Cluster Assembly—Conveying the Milk—Producing Clean Milk—Cooling and Storing Milk—Maintenance and Testing—The Machine and Milking Efficiency—Milking Machines and Mastitis

ILLUSTRATIONS

The photographs were taken by the authors, with the exception of the following suppliers: Alfa-Laval Ltd (Plate 21); D.S.W. Cooney (Plates 1, 4, 7, 10, 16, 22, 23, 34, 36, 38, 39, 40); Farmhand (UK) Ltd (Plate 14); Gascoigne, Gush & Dent Ltd (Plates 19, 28); A. Kidd Ltd (Plate 13); T. C. McCreath (Plate 12); the National Institute for Research in Dairying (Plates 35, 43, 44); and N. H. Strachan (Plates 2, 3, 17, 11, 26, 33)

TABLES

FIGURES

FOREWORD

The modern dairy cow is one of the most fascinating of man's domestic animals, and dairy farming is a major part of British and indeed of Western agriculture.

The cow, 'the foster mother of the human race', produces a foodstuff of high nutritional value which as milk or milk products is an important constituent of Western diets. Moreover it can use feeds such as grass, forage crops and by-products which cannot be used directly for mankind. Because of its complement of symbiotic microorganisms, its biological efficiency and its capacity to use forage, the modern dairy cow is one of the most valuable of our domesticated animals in converting feeds to food for man. Well-managed dairy herds in Britain can produce some 140 kg of edible protein and 11,000 MJ of edible energy from each hectare of land devoted to producing their food. Although chickens can produce similar protein yields, they depend almost entirely on concentrated foods. While meat production from breeding herds of cattle or breeding flocks of sheep produces only about 35 kg of protein and 3,500 MJ of edible energy per hectare, when beef production is a by-product of milk production the combined yields of milk and meat can reach 120 kg edible protein and 10,000 MJ edible energy per hectare.

The high productivity of the dairy enterprise is a function both of the inherent efficiency of the cow and of the application of the knowledge which has been accumulated as a result of research on dairy production in Britain. Our knowledge of dairy cattle breeding, feeding, milking, management and disease control has increased rapidly in recent years. As a result, dairy farming which, like so many other aspects of agriculture, was until recently more an art than a science, is now securely founded on a technological base.

Moreover, in the United Kingdom and in many other countries dairy farming is now a particularly well-organized industry, with an effective dairy farming infrastructure for marketing, information and the provision of technical advice.

For these reasons, dairy farming has for long been regarded as one of the most reliable and rewarding methods of farming, repaying both in intellectual interest and in financial terms the undoubtedly heavy and continuous demands which it makes on dairy farmers and their workers.

To keep abreast of such a comprehensive technology and such a vigorous industry is not easy, and a publication which embraces many recent developments is to be welcomed. The present book prepared by a prominent research worker and an equally prominent farmer, consultant and journalist, does much to assist the farmer or student who wishes to keep up to date. Without going into overmuch technical detail, this book outlines current developments and thinking, not only on the more esoteric aspects of energy metabolism, reproductive physiology or animal behaviour but on such strictly practical activities as grazing management, building layout, slurry disposal and business management on the dairy farm. The book contains a wealth of knowledge and experience gained by the authors, who are both involved in the practical day-to-day management of sizeable dairy herds. It should inform the student, stimulate the farmer and encourage both to study even further the complex and fascinating central character of the book, the dairy cow.

November 1977

W. Holmes
Wye College, Kent

PREFACE TO THE FIRST EDITION

The original idea for producing this book arose from a conversation between one of the authors and Mr R. J. Halley, Deputy Principal of Seale-Hayne Agricultural College, Devon. For this, and for his subsequent assistance in the planning of the subject matter and in reading and commenting on the typescript, we are most grateful.

In writing this book, we have drawn widely and deeply on the opinions, ideas and experience of many friends and colleagues in farming, research, education and the advisory services. We are especially grateful to Professor J. A. F. Rook, Director of the Hannah Research Institute, Ayr, and to the Council of the Institute, for their constant support and encouragement, and to Professor W. Holmes, Wye College, Kent, for writing the Foreword and for suggesting changes in the text.

We would particularly thank the following people who have read chapters and offered constructive comments: Dr D. Reid, Dr P. C. Thomas and Mr J. N. Watson, Hannah Research Institute; Mr G. K. R. Black, Mr W. D. Carr, Dr R. D. Harkess, Dr R. Hissett, Mr R. Laird, Dr J. D. Leaver and Mr W. G. W. Paterson, West of Scotland Agricultural College; Mrs M. E. Vale, Mr R. Y. Macpherson, Mr W. S. Shattock and Mr J. D. Young, ADAS; Dr R. J. Esslement, Reading University; Mr G. F. Smith, MRCVS, Milk Marketing Board; and Mr I. A. MacMillan, MRCVS, Ayrshire.

The editors and publishers of *Agricultural Progress* and *Animal Production* have generously allowed us to reproduce their material in Fig. 6.1, and Figs 8.3 and 17.2 respectively, and we thank them sincerely. *Animal Nutrition*, by P. McDonald, R. A. Edwards and J. F. D. Greenhalgh, has been a constant source of reference, and we thank the authors for permission to quote from their Tables. In Chapter Eleven we were helped by Mr A. J. Quick, ADAS, and are grateful to him and to the editor of *Better Management* for permission to reproduce Table 11.3. Data from the Milk Marketing Boards have formed the basis of many Tables, and we are particularly indebted to Dr J. A. Craven and Mr W. A. Herbert for assistance. The Tables on food composition and ME allowances are reproduced from Technical Bulletin 33 (1975) with kind permission of the Controller of Her

Majesty's Stationery Office. Fig. 10.5. is reproduced with permission of the British Standards Institution.

The photographs were largely the responsibility of Mr N. H. Strachan and Mr D. S. W. Cooney, Hannah Research Institute, and their skill and patience is acknowledged with gratitude. We are indebted also to the following for providing photographs: The National Institute for Research in Dairying, Reading; Alfa-Laval Ltd, Cwmbran; *Dairy Farmer*, Ipswich; Farmhand (UK) Ltd, Wymondham; Gascoigne, Gush & Dent Ltd, Reading; A. Kidd Ltd, Devizes; and T. C. McCreath, Garlieston. We are also grateful to Miss E. Macdaid for calculating the county milk data, and for producing Fig 1.1. Special thanks go to Mrs A. Waldron for her expert typing, and to Mr Neil Hyslop, for his care and accuracy in redrawing the figures.

Finally we both thank our wives for their help, understanding and encouragement throughout the writing of this book.

Malcolm E. Castle   Paul Watkins
Hannah Research Institute,   Pastures Farm,
Ayr, Scotland   Sotherton, Halesworth
                                         Suffolk, England

September 1977

PREFACE TO THE SECOND EDITION

The first edition of *Modern Milk Production* has been read widely at home and overseas. However, the technology of milk production continues to advance with rapidity because of a combination of research findings and their development on the progressive dairy farm. Recent examples include the flat-rate system of rationing concentrates, big-bale silage, the new method of allocating protein in the ration and changes in the design of milking parlours. These changes are all reflected in the new edition, together with other developments that have taken place.

Many people have offered comments and constructive criticism during the revision of the book and we are most grateful for their interest. In particular we thank members of staff at the Hannah Research Institute, the West of Scotland Agricultural College, ADAS and the Milk Marketing Boards. Having incorporated the helpful suggestions of many friends and colleagues, we believe that *Modern Milk Production* will continue to be acceptable to both farmer and student.

Malcolm E. Castle   Paul Watkins

June 1983

CHAPTER ONE

The Dairy Industry

Milk Producers and Cattle Numbers—Dairy Breeds—Milk Yields and Composition—Beef and Milk—Milk Marketing—Price Structure

Milk production is the largest single enterprise in British agriculture, with an annual output valued at over £2,000 million. This represents about 22 per cent of the total value of all agricultural output, which compares with the 16 per cent contributed by fat cattle and calves, and the 25 per cent contributed by all arable crops.

The UK dairy industry compares favourably with the industries of other countries in the European Economic Community in terms of herd size and milk yield per cow (Table 1.1). This table also shows

Table 1.1 Milk production in the EEC, 1981

Country	Number of dairy cows ('000)	Average herd size	Milk yield (kg/cow)	Fat content (%)	Liquid milk consumption as % of total utilization
Germany	5,438	13	4,540	3·84	13
France	6,907	16	3,680	3·80	10
Italy	3,031	7	3,385	3·49	30
Netherlands	2,407	37	5,105	4·06	7
Belgium	962	18	3,920	3·52	18
Luxembourg	68	24	4,015	3·86	14
United Kingdom	3,295	53	4,910	3·87	47
Eire	1,458	16	3,315	3·49	11
Denmark	1,020	26	4,870	4·24	8
Greece	361	3	1,855	3·60	?
Total/Mean	24,947	15	4,130	3·83	18

(From *EEC dairy facts and figures*, 1982*)

*Details of sources of tables and figures, where not otherwise mentioned, are given in the 'Further reading' section at the end of each chapter.

the exceptionally high proportion of milk in the UK which is used for liquid consumption, generally the most profitable outlet. However, even in the UK 53 per cent of the milk produced goes for the manufacture of dairy products such as butter and cheese (which together account for over three-quarters of the quantity), cream, condensed milk and milk powder. The dairy manufacturing industry in the UK has the advantage of a milk supply which fluctuates much less than that of such countries as Eire and the Netherlands. This comparatively level production (52·3 per cent in the six summer months and 47·7 per cent in the winter in 1981/2) is largely due to the seasonal price structure, which has encouraged the production of winter milk.

The UK is, however, a net importer of milk products, the proportion of these which is home-produced being at present 64 per cent, but increasing. The factors determining how this proportion will change in the future are partly economic and partly political. Because dairy farmers in the other EEC countries have smaller herds, they form a larger proportion of the population than they do in the UK, and hence have more political influence. The demand for liquid milk and milk products per head in the UK is high in relation to other countries, but there is a downward trend in consumption throughout the EEC, although total milk production is rising.

Milk Producers and Cattle Numbers

The number of milk producers increased during the Second World War and reached a peak in 1950 (Table 1.2). Since then the number

Table 1.2   Registered milk producers in the UK

Year	England and Wales	Scotland	Northern Ireland	UK
	(in thousands)			
1939	136·5	8·7	10·3	155·5
1945	158·0	8·7	24·7	191·4
1950	162·0	8·7	25·3	196·0
1960	123·1	8·0	20·5	151·6
1970	80·2	5·4	15·1	100·7
1982	40·2	3·1	8·9	52·2

(From *United Kingdom dairy facts and figures*, 1976 and 1982)

has declined steadily, with a large decline between 1970 and 1982, during which period the number of producers in the UK fell from 100,700 to 52,200. Predictably, the largest decreases have been in the parts of the country which are better suited to other types of farming: the South, South-East and East of England.

In contrast, there has been a slow increase in the number of dairy cows in the UK, with the result that average herd size has increased from 20 cows in 1960 to 62 cows in 1982 (Table 1.3). This figure is still rising at a rate of about 2 cows per year. It is interesting to note the regional variations in herd size: 30 cows in Northern Ireland, 63 cows in England and Wales and 85 cows in Scotland.

Table 1.3 Dairy cow numbers and average herd size in the UK

Year	Number of dairy cows ('000)	Average herd size
1960	3,165	20
1970	3,244	31
1975	3,242	43
1980	3,224	55
1982	3,250	62

(From *United Kingdom dairy facts and figures*, 1982)

It seems probable that the trend towards larger herds will continue as a gradual process, but there seems little likelihood of a general move towards herds of several hundred cows. Two-thirds of all the cows in England and Wales are still in herds of less than 100 cows.

Output of milk per county (Fig. 1.1) shows a wide variation. As might be expected, the counties with the highest production per ha (hectare) are those such as Cheshire, part of Dyfed, Somerset and Dorset which, while situated in the grassier West of the country, are not rendered unsuitable for dairying by quality of land or difficulty of access. On the other hand, production is lower in the Eastern counties, where arable crops can be grown profitably.

Although there are many dairy units on mixed farms, dairy farms on the whole tend to be comparatively small, specialized units on which most of the work is done by the farmer and his family. Unpaid labour accounts for about half of all working hours per cow, and on small dairy farms it is often 80 to 90 per cent of the total.

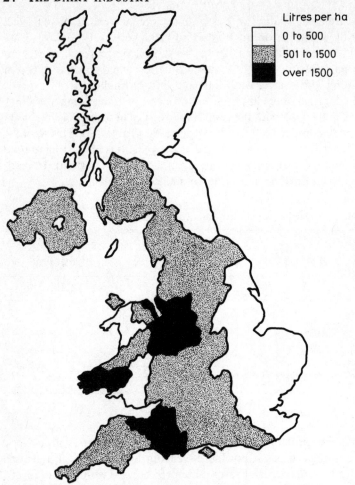

Fig. 1.1. Milk production on a county basis, 1981–2
(Based on *United Kingdom dairy facts and figures*, 1982)

Dairy Breeds

The principal dairy breeds found in the UK are listed in Table 1.4. All have their adherents. The British Friesian is favoured for its high milk production and good beef qualities; the Ayrshire for its relatively low food requirement and good butterfat content; and the Channel Island breeds for the very high fat content of their milk.

The past 20 years have seen a big change in the distribution of the

Table 1.4. Distribution of cows and heifers in milk and cows in calf by breed in the UK

Breed Type	(Percentage of total animals) England and Wales		Scotland		Northern Ireland	
	1973–4	1978–9	1975	1981	1972	1979
Ayrshire	3·6	3·4	47·0	28·8	4·4	1·6
Friesian	81·0	88·6	27·9	46·9	82·4	94·6
Holstein	—	1·4	—	1·0	*	*
Guernsey	2·8	2·4 ⎫	0·7	0·3	*	*
Jersey	2·2	2·0 ⎭			*	*
Dairy Shorthorn	0·9	0·4	*	*	11·6	3·1
Crosses and others	9·5	1·8	24·4	23·0	1·6	0·7
	100·0	100·0	100·0	100·0	100·0	100·0

*Included in 'Crosses and others'
(From *United Kingdom dairy facts and figures*, 1982)

breeds of cow in England and Wales, particularly notable being the rise of the British Friesian to dominance, the decline of the Ayrshire and Guernsey and the virtual disappearance of the Dairy Shorthorn

Table 1.5. Inseminations, by breed, 1981–2

	(Percentage of total) England and Wales	Scotland	Northern Ireland
Dairy			
Ayrshire	1·0	9·4	0·3
Friesian/Holstein	61·7	58·1	43·7
Others	2·3	—	*
Dual purpose			
All breeds	0·5	0·9	1·6
Beef			
Charolais	7·0	6·8	10·9
Hereford	17·7	8·5	13·2
Simmental	0·9	4·3	18·0
Others	8·9	12·0	12·3
	100·0	100·0	100·0

*Included in 'Beef, Others'
(From *United Kingdom dairy facts and figures*, 1982)

(Table 1.4). In Scotland, while the Ayrshire has also declined, it is still an important breed and constitutes 28·8 per cent of all dairy cows. Friesian/Ayrshire crosses contribute 20·1 per cent of all dairy cows in Scotland. There are interesting regional variations, South Wales being dominated by the Friesian, while the Channel Island (Guernsey and Jersey) breeds are popular in South-East England, where there is a good demand for their high-fat milk.

An indication of trends in the popularity of the various breeds is given by the insemination figures of the Milk Marketing Boards (Table 1.5), but these figures do not show the true picture since the proportion of herds using artificial insemination only is 66.1 per cent in England and Wales and as low as 10.1 per cent in Scotland.

Milk Yields and Composition

Throughout this book, 1 litre of milk is considered to be equivalent to 1 kg of milk, although the exact equivalent is 1·03 kg. Both units are used in dairying and for most practical purposes they can be taken as being the same. In most modern milking systems, milk is measured volumetrically, i.e. in litres, rather than weighed—for example, in graduated recording jars, and by flow meter from a bulk tank. Although the quantity of milk is expressed as kg in official milk recording schemes, when milk is sold the amount is stated in litres.

The average annual milk yield of all dairy cows in the UK was 4,745 litres in 1981–2. The yields in Table 1.6 show a steady increase

Table 1.6 Average annual milk yield per cow

| | (litres) | | |
Year	England and Wales	Scotland	Northern Ireland
1964–5	3,545	3,545	3,180
1969–70	3,755	3,750	3,660
1977–8	4,570	4,475	4,360
1981–2	4,785	4,765	4,450

(From *United Kingdom dairy facts and figures*, 1982)

since 1964–5; the rate of increase is being maintained and shows no signs of decreasing. The yields in Table 1.7 are from officially recorded herds, which are generally about 600 litres per cow higher than those of non-recorded herds. Official testing for the crude protein content of milk only started in 1976, and thus data for this important constituent of milk are not as extensive as those for fat.

Table 1.7. Milk yield, fat and crude protein content in recorded herds of different breed types in England and Wales, 1980–1

Breed type	Average yield (kg per cow)	Fat (%)	Crude protein (%)
Ayrshire	4,977	3·94	3·38
British Friesian	5,606	3·83	3·27
British Holstein	6,295	3·77	3·21
Dairy Shorthorn	4,908	3·65	3·32
Guernsey	4,052	4·66	3·62
Jersey	3,879	5·22	3·86
Mean (including other breeds)	5,521	3·86	3·29

(From *United Kingdom dairy facts and figures*, 1982)

The average composition of the milk from all herds in England and Wales in 1981–2 was 3·88 per cent butterfat, 8·71 per cent solids-not-fat, total solids 12·59 per cent.

Beef and Milk

Over half (62 per cent) of the beef produced in the UK comes from the dairy herd, either from the fattening of bull calves and surplus heifer calves (43 per cent) or as beef from cull cows and bulls (19 per cent). The beef by-product is an important aspect of the dairy industry, and the profitability of dairy farming is closely linked to that of beef production. The Friesian breed, with its good beef characteristics, has played an important role in establishing this position. In many dairy herds, a sufficient number of cows are bred pure in order to maintain herd numbers and the remainder are crossed with a beef bull. The beef-cross calves tend to fetch a higher price than surplus pure-bred dairy calves unless the latter have a particularly high value for breeding.

In general, beef production needs a considerably lower capital investment in buildings and equipment than dairy farming, although in intensive systems the investment in stock may be as high. Beef production uses less labour than dairying, but has in the past produced much lower gross margins per ha. Specialist beef production tends to be centred mainly in parts of the country—such as hill and upland—which are not suitable for milk production.

Milk Marketing

The marketing of milk in the UK has been the responsibility of the Milk Marketing Boards since 1933. These Boards are a type of compulsory farmers' co-operative, and operate as independent producers' organizations within legally constituted powers. There are 5 Boards in the UK: 1 in England and Wales, 3 in Scotland, and 1 in Northern Ireland. A farmer offering milk for sale must be registered with the local Board which must, in turn, buy the milk and find a market for it.

At present the Boards are responsible for implementing the pricing policy for milk and also perform a range of marketing and technical services including milk transport, the operation of about 60 creameries, advertising and sales development work. The farm activities of the Boards include the operation of an artificial insemination service, a milk-recording service, and management accounting and advisory schemes. The artificial insemination service has 120 centres throughout the UK with about 1,000 bulls of 15 different breeds. Over half the dairy cows in the country are now bred by artificial insemination.

Price Structure

The net price for milk received by producers averaged 13·7p per litre in 1981–2, with some small differences between the prices paid by the five Boards. The price received by producers in the various months shows a distinct seasonal pattern, which is designed to ensure as level a supply of milk as possible throughout the year. Current management practices result in a flush of milk in early summer and a shortage in late summer, and the price schedule aims to correct this.

The price of milk is fixed annually under regulations within the European Economic Community, and the maintenance of the price is achieved by setting intervention prices for certain milk products.

In principle, trade in milk between member states is possible if health regulations permit.

Further Reading

An analysis of FMS costed farms 1981/82, Report No. 33, 1982, Milk Marketing Board, Reading

Beef yearbook, 1983, Meat and Livestock Commission, Bletchley, Milton Keynes

EEC dairy facts and figures, 1982, Milk Marketing Board, Thames Ditton, Surrey

Report of the breeding and production organisation. No. 32, 1981–82, Milk Marketing Board, Thames Ditton, Surrey

The structure of Scottish milk production at 1981, 1981, Scottish Milk Marketing Board, Paisley

United Kingdom dairy facts and figures, 1982, Federation of United Kingdom Milk Marketing Boards, Thames Ditton, Surrey

CHAPTER TWO

Feeling Dairy Cows

Food Constituents—Organic Matter—Digestion—Digestibility—Products of Digestion—Utilization of Digested Products—Food Energy—Dry-matter Intake—Indigestible Organic Matter—Liveweight Changes—New Protein System—Feeding—Allocation of Concentrates—Minerals—Vitamins—Relative Feed Costs

The feeding of dairy cows is without doubt an art, but for success in this skilful operation it is important to know and to practise some well-established scientific principles in animal nutrition. A vital feature in the correct feeding of dairy cows is to know the requirements of the cow for specific nutrients, the properties and feeding value of the common foods, and finally how to bring these two aspects of feeding together in a profitable system of milk production. Food, either purchased or home-grown, accounts for over 60 per cent of the total costs of producing milk, and it is thus extremely important to be efficient in feeding the dairy cow.

Food Constituents

The main components of foods are shown in the following diagram.

```
                ┌ Water                                    ┌ Proteins
                │                                          │ Oils
Food ──────────┤                  ┌ Organic matter ──────┤ Fibre
                │                  │                       │ Carbohydrates
                └ Dry matter ─────┤                        └ Vitamins
                                   └ Inorganic matter──────Minerals
```

Almost all foods contain water, and when the food is heated at 100 °C to a constant weight, the water is evaporated and the dry matter remains. The proportion remaining varies between different foods, and can range from 85·0 per cent in barley to 8·5 per cent in turnips (Table 2.1). The importance of knowing the dry-matter content of foods cannot be overemphasized, especially in the making and feeding of silages. At one extreme, grass for silage making may contain only 16·0 per cent dry matter and produce a low dry-matter silage, whereas wilted herbage may make a silage with 32·0 per cent dry matter. The water in foods is a normal source of water for the cow, and if the ration contains a high proportion of succulent foods such as turnips and silage, the intake of drinking water is usually markedly reduced. Cows which are not milking, and which eat large amounts of silage, may hardly drink at all for periods of many weeks.

Table 2.1. Proximate composition of some selected foods for dairy cows

Food	Dry matter	Crude protein	Ether extract	Crude fibre	Nitrogen-free extractives	Ash
			(*Percentage of total food*)			
Turnips	8·5	1·0	0·2	0·9	5·7	0·7
Barley	85·0	9·0	1·5	4·5	67·4	2·6
Soya bean meal	90·0	45·3	1·5	5·2	32·4	5·6
Grass silage	20·0	3·6	1·0	5·3	8·1	2·0
			(*Percentage of dry matter*)			
Turnips		11·8	2·4	10·6	67·0	8·2
Barley		10·6	1·8	5·3	79·3	3·0
Soya bean meal		50·3	1·7	5·8	36·0	6·2
Grass silage		18·0	5·0	20·5	46·5	10·0

(Based on *Energy allowances and feeding systems for ruminants*)

The dry matter in foods consists of organic matter, which is burnt away if the food is heated at 500 °C, and inorganic matter, termed minerals or ash, which remains after heating. The ash contains silica, calcium, phosphorus, magnesium, sodium, potassium and trace elements which are vital to the health and well-being of the cow. The amount of minerals in foods varies widely. Cereals such as barley are relatively high in phosphorus but low in calcium, whereas silage can be low in phosphorus and high in calcium.

Organic Matter

The organic matter of the food contains the proteins, oils, fibre, carbohydrates and vitamins. The *crude protein* content is calculated from the total nitrogen in the food and contains the true protein, the amino acids which form the proteins, and other non-protein nitrogen compounds. Foods should not be assessed merely on their crude protein contents, but also on the basis of the proportion of nitrogen in the feed which is broken down in the rumen, i.e. the rumen-degradable protein. Some grass silages may have a relatively high content of crude protein, but 60 per cent of the total nitrogen can be in the form of non-protein nitrogen, and only 20 per cent of the original protein in the grass passes intact beyond the rumen in the undegraded form (Table 2.6).

The oil content, correctly termed *ether extract* because of the method of estimation, consists of fats, waxes and oils, and is an important source of energy for the dairy cow. The oil content in many farm foods is low (Table 2.1), and fats and oils are generally added to compound concentrates (Chapter Seven).

The *crude fibre* in foods consists of cellulose, hemi-cellulose, lignin and other structural parts of the foodstuff. The content of crude fibre is a measure of the less digestible part of the food and is useful in valuing foods such as hay. However, it is known that a large part of the crude fibre can be digested and provide energy for the cow.

The readily available carbohydrates in the food are mainly provided by the sugars, starches and other components in a fraction termed the *nitrogen-free extractives*. Carbohydrates can be a large fraction of the food, e.g. in barley (Table 2.1), and are an important source of energy for the cow. The *vitamins* are present in small amounts, but can have far-reaching effects on the health and performance of the cattle.

When foods are analysed, the system termed 'proximate analysis' divides the food into the six fractions (Table 2.1) already described, i.e. dry matter, crude protein, ether extract, crude fibre, nitrogen-free extractives and ash (minerals). An alternative method of giving the same data is to express the fractions as a percentage of the dry matter, which in many situations is an improved method (Table 2.1).

Digestion

Food is digested in the alimentary tract, which is a tube extending from mouth to anus along which the food passes (Fig. 2.1). In the cow, food travels down the oesophagus to the rumen, which has a capacity of about 130 litres. After regurgitation and rumination, i.e.

cudding, the food re-enters either the rumen or the reticulum before passing to the omasum and then the abomasum. The 4 compartments of the cow's stomach are the rumen, reticulum, omasum and abomasum. Food is mixed with large volumes of saliva when it is eaten and ruminated, and the contents of the rumen contain about 90 per cent water and 10 per cent solid material.

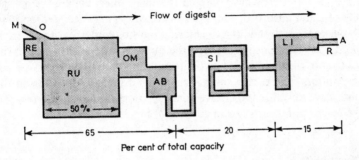

Fig. 2.1. Diagrammatic representation of the digestive tract of the cow Key: M = mouth; O = oesphagus; RE = reticulum; RU = rumen; OM = omasum; AB = abomasum; SI = small intestine; LI = large intestine; R = rectum; A = anus
(Adapted from R. D. Frandson, *Anatomy and Physiology of Farm Animals*, 1975, Lea and Febiger, Philadelphia)

The chemical breakdown of the food in the rumen and reticulum is effected by enzymes secreted by the hosts of bacteria and protozoa which are present in these parts of the digestive tract. There are about 10,000 million bacteria and 1 million protozoa per ml of rumen contents; the surface area of all the organisms in the rumen of a cow is about 0·5 ha. The activity of these organisms within the rumen is immense, and 60 to 70 per cent of all digestible dry matter entering the rumen is converted into compounds which are directly absorbed from the rumen into the bloodstream. The other 30 to 40 per cent continues onwards through the alimentary tract (Fig. 2.1) via the small intestine and large intestine where the food is further digested and absorbed, until the solid waste material, i.e. the faeces, is finally excreted.

The rumen is the distinctive, and vitally important, part of the digestive tract, and the term 'ruminants' is used to describe cattle and other animals with a rumen. The diet of the cow normally contains much forage either eaten directly as pasture or in the conserved form such as hay and silage. All these foods contain relatively large amounts of crude fibre (Table 2.1) and numerous nitrogen-free extractives such as sugars and starch, which are attacked by the rumen micro-organisms and subsequently digested in the rumen.

The protein in the food is also broken down by the bacteria and protozoa in the rumen to simpler substances such as amino acids and ammonia. These organisms build up protein within their own bodies and thus, after they have entered the small intestine, are themselves digested and absorbed. This microbial protein may account for 50 to 90 per cent of all the protein entering the small intestine and there is often little relationship between the quantity of protein eaten in the food and the amount of protein entering the small intestine for absorption. The rumen micro-organisms dominate digestion in the ruminant, and it is imperative that the rumen functions correctly if the cow is to be healthy and capable of high production.

Digestibility

Although the analysis of a food may be a useful guide to its feeding value, the real value of that food to the cow depends on the amount which is digested by the animal. Digestibility is defined simply as the proportion of the food which is absorbed by the animal and which is not excreted and wasted in the faeces. For example, if a cow consumed 50 kg grass silage per day containing 10 kg dry matter, and excreted 4 kg dry matter in the faeces per day, then by difference 6 kg dry matter was absorbed and the digestibility of the dry matter, expressed as a percentage, would be $(10-4) \div 10 \times 100 = 60$ per cent. This value is termed the apparent digestibility coefficient. Digestibility coefficients for protein and the other constituents of the food are calculated in an identical way.

A widely used measure of digestibility in forages is the content of digestible organic matter in the dry matter, which is termed either DOMD, or more simply the D-value. This value is extremely useful in describing and assessing foods and is derived by multiplying the digestibility coefficient of the organic matter by the content of organic matter in the food dry matter. The D-value of a poor hay may be about 50, whereas that of an excellent silage may be 70.

The digestibility of foods is normally measured by giving animals exact amounts of food, and accurately recording the weight of faeces. This measurement is termed an *in vivo* determination, which means in the living animal. An alternative method of measuring digestibility uses small amounts of rumen contents and food in a type of test tube. This is termed an *in vitro* technique, which means in glass. This latter technique has many valuable uses in evaluating foods.

The digestibility of foods varies widely and can be altered by many factors. For example, the digestibility of young spring grass is high, 70 to 75 per cent, but decreases as the grass becomes more mature and fibrous at a rate of about 3 per cent per week. The digestibility of cereal straws is low but can be increased by treatment with an alkali such as sodium hydroxide (Chapter Seven). Grinding a forage will reduce the digestibility by 5 to 15 per cent since the fine material passes through the rumen more quickly than either long or chopped material. In general as an animal is given more food, the digestibility falls since the material passes more quickly through the digestive tract.

Products of Digestion

An outline summary of the main pathways of digestion in the ruminant is given in Fig. 2.2. Fibrous and starchy carbohydrates, the main constituents of energy in the diet, are fermented in the rumen to form short-chain fatty acids. Depending on the type of ration, the mixture of acids consists of 40 to 70 per cent acetic acid, 15 to 40 per cent propionic acid and 10 to 30 per cent butyric acid. Smaller amounts of iso-butyric acid, n-valeric acid and iso-valeric acid plus methane are also produced.

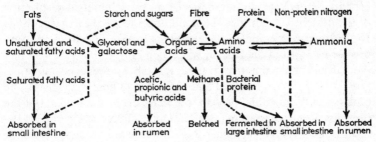

Fig. 2.2. Outline of the main pathways of the digestion of food constituents in the cow
Note: dotted lines indicate an incomplete fermentation in the rumen
(After P. C. Thomas, 'Diet and milk secretion in the ruminant', *World Review of Animal Production*, 11, 1965, p. 33)

A ration of long forage and a small amount of concentrates will be associated with a high proportion of acetic acid, 65 to 70 per cent, a low proportion of propionic acid, 15 to 20 per cent, and a variable proportion of butyric acid, 8 to 15 per cent. Increasing the level of concentrate feeding will usually lower the proportion of acetic acid and increase either propionic or butyric acid, or both. The acids are absorbed across the rumen wall and utilized in the tissues, whereas the methane is belched and thus wasted as a source of energy for the cow. In the rumen, 75 to 80 per cent of the energy from the food is absorbed in the form of volatile fatty acids.

The sugars and starches in cereals are virtually all fermented in the rumen, and only 5 to 10 per cent of the starch in the ration enters the small intestine where it is converted to simple sugars and absorbed. The fibrous food components such as cellulose and hemi-cellulose are not completely digested in the rumen, and a substantial part of the digestible fibre passes to the abomasum. Fibre is not attacked by the enzymes in the small intestine, and the digestion is largely confined to the large intestine, which is a second site of microbial fermentation.

The fats in the ration, mainly in the form of triglycerides, are modified in the rumen to form glycerol, galactose and long-chain fatty acids, of both the saturated and unsaturated type. The glycerol and galactose are fermented to short-chain fatty acids, but the long-chain saturated fatty acids are unattacked and pass to the small intestine, where they are absorbed into the blood and the lymph.

Food proteins are converted in the rumen to amino acids which in turn are changed to fatty acids and ammonia. The breakdown of the protein, i.e. its degradation, depends on the solubility and structure of the protein, but for most common farm foods the degradation is 70 to 80 per cent complete. Ammonia is formed also from the non-protein nitrogen in the rumen, which comes from both the diet and the saliva. The ammonia is used for the synthesis of microbial protein, and is also absorbed and converted to urea in the liver. The production of microbial protein is a vital part of the function of the rumen, and the presence of fatty acids and sulphur is essential for the process. On average, 30 g microbial nitrogen are produced per 1 kg organic matter apparently digested in the rumen. The microbial protein and any unfermented dietary protein is digested in the abomasum and small intestine, where the liberated amino acids are absorbed.

Utilization of Digested Products

Digested nutrients are utilized for the maintenance processes of preserving and replacing body tissue, and for the productive processes of growth and milk production. In mature animals, growth is a small item, but there are stages in the lactation when fat is either deposited or utilized in the body.

Milk is synthesized in the alveolar cells (Chapter Ten) of the mammary gland from precursor substances in the blood by a series of complex processes (Table 2.2). Some of the constituents of milk are transported unchanged from the blood, whereas other constituents are unique products made only by mammary cells. On average, 90 per cent of the protein, 50 to 60 per cent of the fatty acids, and all the lactose are synthesized from precursors absorbed from the blood, and the remainder originates from the plasma. To maintain the supply of precursors to the mammary cells of a cow yielding 20 kg milk per day a minimum of 6 kg of blood per minute flows through the udder; the equivalent of 9 t (tonnes) per 24 hours.

Table 2.2. Relationships between milk constituents, their precursors in the blood, and the end-products of digestion in the cow

Milk constituent	Main precursor in blood	End-product of digestion	Site of absorption
Lactose	Glucose	(1) Propionic and lactic acid	Rumen
		(2) Glucose and amino acids	Small intestine
Protein	Amino acids	(1) Propionic and lactic acid	Rumen
		(2) Glucose, amino acids	Small intestine
Fat	Acetate β-hydroxy-butyrate	(1) Acetic and butyric acid	Rumen
	Triglycerides	(2) Long-chain fatty acids	Small intestine
	Glucose	(3) Propionic and lactic acid	Rumen
		(4) Glucose and amino acids	Small intestine

(After Armstrong, D. G., and Prescott, J. H. D., I. R. Falconer (Ed.), *Lactation*, 1971, Butterworths, London)

The milk precursors may arise directly from the digestive tract, partly as modified products of digestion after passing through the liver, and partly after metabolic changes within the liver tissue. For example, acetic acid absorbed from the rumen passes through the liver with relatively little change, whereas in contrast, butyric acid is changed in the rumen wall and liver to β-hydroxybutyrate and other ketone bodies. Part of the propionic acid is changed in the rumen wall to lactic acid and utilized in the liver to form glucose. The main precursors of milk are thus acetic acid, β-hydroxybutyrate, glucose, amino acids and long-chain fatty acids.

Lactose in milk is synthesized mainly from blood glucose, a deficiency of which will reduce the amount of milk secreted. Milk proteins are synthesized mainly from free amino acids in the blood. Milk fat arises from the fatty acids of the blood triglycerides or is synthesized from acetate and β-hydroxybutyrate of the blood plasma. Changes in rations which alter the proportions of volatile fatty acids in the rumen will affect the amount of fat in the milk, whereas the feeding of different fats will alter the proportion of short-and long-chain fatty acids in the milk fat.

Food Energy

An important function of food is to supply energy to the animal, and thus the energy value of food is of vital importance. For many years the contents of the digestible nutrients in a food were used to calculate a starch equivalent value, but this system of feed evaluation and feed requirements has now been changed to a system based on metabolizable energy (ME). The ME value of a food may be determined in an animal feeding experiment using a respiration chamber or calorimeter. The method is similar to a digestibility trial, and the faeces, urine and expired gases, which include methane, are all collected. The content of energy in the food is determined, and from this value is deducted the energy value of the faeces, urine and methane. The resulting value is the ME of the food, which is expressed as megajoules (MJ) per kg dry matter (DM), as shown in the following example, which is of grass given to sheep.

$$
\begin{array}{llll}
\text{Dry-matter intake per day} & & & = 2 \cdot 0 \text{ kg} \\
\text{Gross energy intake} & & & = 35 \cdot 0 \text{ MJ} \\
\quad \text{Energy in faeces} & = 14 \text{ MJ} \\
\quad \text{Energy in urine} & = 1 \text{ MJ} \\
\quad \text{Energy in methane} & = 2 \text{ MJ} \\
\quad \quad \text{Total} & = & & 17 \cdot 0 \text{ MJ}
\end{array}
$$

$$
\text{Metabolizable energy} = \frac{35 \cdot 0 - 17 \cdot 0}{2 \cdot 0} = 9 \cdot 0 \text{ MJ per kg DM}
$$

Simpler, but less accurate, methods of calculating the ME values of foods involve the use of equations and the analysis of the food. Some typical ME values are given in Appendix 1.

The animal's need for energy is also measured in a calorimeter, and for dairy cows the system involves a separate calculation of the allowances for maintenance and for milk production. These two allowances are added to give a total ME allowance which can be altered for changes in liveweight. The daily maintenance allowances of ME for dairy cows of different weights are given in Table 2.3.

Table 2.3. Daily maintenance allowances of ME, digestible crude protein (DCP), tissue protein (TP) and the indigestible organic matter (IOM) appetite values for dairy cows

Breed	Liveweight (kg)	MJ/day	DCP* (g/day)	TP (g/day)	IOM (kg/day)
Dexter	300	36	210	56	1·8
Jersey	350	40	230	60	2·1
Kerry	400	45	250	64	2·4
Guernsey	450	49	270	67	2·7
Ayrshire	500	54	290	69	3·0
Shorthorn	550	59	310	72	3·3
Friesian	600	63	330	74	3·6
South Devon	650	67	350	76	3·9

*Plus 270 g DCP per day for pregnancy in last 2 months before calving
(From *Energy allowances and feeding systems for ruminants, Nutrient allowances and composition of feedingstuffs for ruminants* and *The nutrient requirements of ruminant livestock*)

If the cows are pregnant, the allowances should be increased. For example, a Friesian cow weighing 600 kg has an allowance of 63 MJ per day (Table 2.3), which rises to 71 MJ in the sixth month of pregnancy and to 83 MJ in the ninth month. Up to the fifth month of pregnancy the extra allowance is generally less than 5 MJ per day.

The ME allowances for milk production are shown in Table 2.4. If the exact fat and solids-not-fat (SNF) content of the milk is known, an even more precise allowance of MJ per kg milk can be calculated, but for milk of average composition a value of 5 MJ per kg milk is sufficiently accurate.

Table 2.4. ME, DCP and TP allowances per 1 kg milk

Breed	Fat content of milk (%)	MJ/kg milk	DCP (g/kg milk)	TP (g/kg milk)
Friesian	3·5	4·9	51	30
Ayrshire	3·7	5·1	53	31
High-fat breeds	4·8	5·9	67	36

(From *Energy allowances and feeding systems for ruminants*, *Nutrient allowances and composition of feedingstuffs for ruminants* and *The nutrient requirements of ruminant livestock*)

If a cow is gaining or losing liveweight, a further allowance should be made. For 1 kg liveweight gain, an allowance of 34 MJ is required, whereas for 1 kg of liveweight loss there has been a contribution from the body tissue equivalent to 28 MJ of dietary ME. An example of a Friesian cow weighing 600 kg, producing 20 kg milk per day 2 to 3 months after calving and gaining in liveweight illustrates the system of allowances.

	MJ per day
Maintenance 600 kg (Table 2.3)	63
Milk, 20 kg per day @ 4·9 MJ per kg (Table 2.4)	98
Weight gain, 0·5 kg per day @ 34 MJ per kg	17
Total	178

The total ration of the cow should supply 178 MJ per day.

Dry-matter Intake

Before calculating a ration for a cow an estimate of the dry-matter intake of the animal must be made. A general guide is to allow 2·5 to 3·0 kg dry matter per day per 100 kg liveweight, but factors such as milk yield, stage of lactation, type of ration and method of feeding also affect intake, and allowance must be made for them in determining the ration.

At calving, the dry-matter intake of cows is normally at a minimum, with a progressive increase to a maximum at 4 to 5 months after

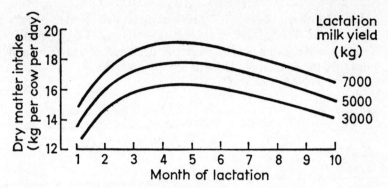

Fig. 2.3. Some predicted maximum intakes of dry matter for lactating Friesian cows weighing 600 kg
(After J. F. D. Greenhalgh and I. McDonald, 'The ME system in practice —predicting feed intake', *Animal Production*, 24, 1977, p. 136)

calving (Figs. 2.3 and 2.4). After this peak, the intake declines slowly as the lactation advances. A further indication of the large variation in voluntary intake at different stages of the lactation is illustrated in the following estimates of the percentage values of the mean intake (100 per cent) in the successive months of lactation.

Month	1	2	3	4	5	6	7	8	9	10
Percentages	81	98	107	108	109	108	101	99	97	93

The approximate dry-matter intake for cows in mid and late lactation can be calculated from the following equation in *Energy allowances and feeding systems for ruminants* ('Further Reading' at end of this Chapter).

Dry-matter intake (kg per day) = 0·025 × liveweight (kg) + 0·1 × milk yield (kg per day)

The following example illustrates this calculation for a Friesian cow.

		kg dry matter per day
Liveweight, 600 kg × 0·025	=	15·0
Milk, 20 kg per day × 0·1	=	2·0
Total		17·0

In early lactation, this intake would be lower by about 2 to 3 kg dry matter per day. To increase nutrient intake, foods of the highest digestibility should be given; an increase in the number of separate feeds may also increase intake.

In the two examples given previously the cow needed a daily allowance of 178 MJ and a dry-matter intake of 17·0 kg per day. The following ration would meet the demands of the cow, and illustrates the method of calculation:

			Intake per day		
Food	Dry-matter content (%)	MJ/kg dry matter	Food (kg)	Dry matter (kg)	ME (MJ)
Silage	20·0	9·0	39	7·8	70
Barley	85·0	12·9	10	8·4	108
				16·2	178

In this ration, 178 MJ are supplied in 16·2 kg dry matter, and the ME concentration of the entire diet, M/D, is calculated as 178 ÷ 16·2 = 11·0. This measure of energy concentration is a useful additional indication of the suitability of a specific diet at a particular stage of lactation, and some examples are given in Table 2.5. In early lactation, when dry-matter intake is not at its maximum, the highest M/D of 11·5 is required. In late lactation, when the milk yield is 10 kg

Table 2.5. Some intakes of dry matter and ME at different stages of lactation

Weeks after calving	Milk Yield (kg/cow per day)	Dry-matter intake (kg/cow per day)	ME (MJ/cow per day)	M/D (MJ per kg dry matter)
0–10	20	15·0	172	11·5
10–20	25	17·5	186	10·6
20–30	20	17·0	177	10·4
30–40	15	16·5	152	9·2
40–50	10	16·0	135	8·4

(From *Energy allowances and feeding systems for ruminants*)

per day the M/D is 8·4. In Table 2.5 the dairy concentrate was assumed to be given at a rate of almost 5 kg per 10 kg milk in early lactation, and about 4 kg per 10 kg milk in late lactation.

It is emphasized that any system, even the ME system, is only a guide to feeding, and alterations in the ration will have to be made after the production of the cow is observed. Such changes are part of the art of feeding cows, and should be made fairly slowly so that the cow is never upset by a drastic change in the ration. Changes in rations generally change the flora in the rumen, and hence alterations in the diet should never be made suddenly. For example, a change from a winter diet to spring grazing should be made over a period of about 1 week.

Indigestible Organic Matter

An alternative system to that of estimating dry-matter intake when devising rations is the calculation of the amount of indigestible organic matter (IOM) in the ration. The IOM per cent is a measure of the organic matter in the feed which is not digested and is derived from the following equation.

$$IOM \text{ (per cent)} = 100 - (D\text{-value} + \text{per cent ash})$$

Barley has an IOM value of 11 per cent, whereas grass hay of poor quality has a value of 42 per cent. The values for other foods are given in *Nutrient allowances and composition of feedingstuffs for ruminants* ('Further Reading' at end of this chapter). IOM appetite values refer to the total daily ration of the cow, and are listed in Table 2.3. The IOM method allows a suitable ration to be rapidly devised, but the system does not yet allow for reduced intake in early lactation and other factors which influence voluntary intake. If accurate corrections could be made to the basic values given in Table 2.3, the system would have a valuable and wider application.

Liveweight Changes

Changes in liveweight are a normal part of the milk production pattern. At calving, a Friesian cow may lose a total of about 65 kg liveweight, which consists of the calf, weighing 45 kg, and fluids and tissues weighing about 20 kg. The cow invariably loses more live-weight during early lactation as her reserves of fat are depleted in order to sustain a milk yield higher than that which can be supported by her food intake. If the loss of liveweight is too high there may be problems with infertility, a low SNF content in the milk and a reduced peak milk yield.

The loss in liveweight in early lactation should be kept below 0·5 kg per day, i.e. a maximum loss of 35 kg in the first 10 weeks. This level of weight loss is almost inevitable when the intake of dry matter is restricted by appetite (Fig. 2.3) and yet the demands for energy for milk production are high. From weeks 10 to 20 the intake of dry matter increases and the liveweight of the cow should remain fairly static. After week 20, gains in liveweight of 0·5 to 0·6 kg per day are desirable in a well-fed and high-yielding cow.

These figures are only a guide, and a skilled stockman will rapidly detect changes in liveweight and the condition of the cows, and make changes in the ration. Routine weighing of cows is unlikely to become a general farm practice, but a system of 'body scoring' can be most useful (Chapter Fourteen). This technique is a simple method of assessing body condition which indirectly indicates changes in weight and assesses the state of body reserves.

New Protein System

Although energy must receive first priority in devising a feeding system, it is equally important that the amount of protein is adequate for the requirements of the animal. The system used widely and reasonably successfully in the past was one based on digestible crude protein (DCP), and the allowances for maintenance and milk production are given in Tables 2.3 and 2.4 respectively. The DCP content of the main foods is shown in Appendix One.

However, a more dynamic system of protein feeding is now suggested which relates to the supply of both protein and energy, and allows for the protein which is not broken down in the rumen. The system starts by calculating the energy requirements of the cow as shown previously (p. 40) using the ME allowances in Tables 2.3 and 2.4.

The *tissue protein* (TP) requirements of the animal for maintenance, production and liveweight change are then calculated using the TP values in Tables 2.3 and 2.4. The requirements of TP for liveweight gain and loss are 150 and 112 g per kg respectively. An example for a Friesian cow illustrates the system.

	TP (g per day)
Maintenance, 600 kg (Table 2.3)	74
Milk, 20 kg per day @	
30 g per kg (Table 2.4)	600
Weight gain, 0·5 kg per day	
@ 150 g per kg	75
Total	749

To meet this TP requirement, there is a demand for both rumen-degradable protein and undegraded dietary protein.

The *rumen-degradable protein* (RDP) which is required for optimal microbial activity and microbial-protein production in the rumen depends on total energy intake and is calculated as follows.

$$RDP (g/day) = 7·8 \times ME \text{ requirement}$$

Thus from the ME requirement (p. 40), the RDP requirement is $7·8 \times 178 = 1,388$ g per day.

The *undegraded dietary protein* (UDP) is calculated as follows.

$$UDP (g/day) = (1·91 \times TP \text{ requirement}) - (6·25 \times ME \text{ requirement})$$

Using the previous example, the UDP requirement is $(1·91 \times 749) - (6·25 \times 178) = 319$ g per day. Having calculated the TP, and how this can be met by the RDP and UDP, a ration can be devised to meet the requirements for both protein and energy.

The ME content of the main foods are given in Appendix One, and a suggested grouping of foods based on protein degradability is presented in Table 2.6. The RDP value of a food is calculated by multiplying the protein degradability by the content of crude protein. For example, silage containing 12 per cent CP in the dry matter and with a degradability of 80 per cent (Table 2.6) gives an RDP value of $12 \times \dfrac{80}{100} = 9·6$ or 96 g per kg DM. The UDP is calculated by deducting the RDP value from the CP value, i.e. $120 - 96 = 24$ g per kg DM.

Thus the requirements of the 600 kg Friesian cow may be more than met as follows:

	Intake per day			
Food	Dry matter (kg)	ME (MJ)	RDP (g)	UDP (g)
Silage	7·8	70	749	187
Barley	8·4	108	722	185
	16·2	178	1,471	372
Requirement			1,388	319

A more complex ration can be formulated for a 600 kg cow yielding 30 kg milk per day containing 3·7 per cent fat and with no change in liveweight. (From *The nutrient requirements of ruminant livestock*)

	Intake per day			
Food	Dry matter (kg)	ME (MJ)	RDP (g)	UDP (g)
Hay	7·5	63	510	128
Silage	3·2	33	435	109
Barley	3·2	44	275	70
Flaked maize	4·1	62	271	180
Soya bean meal	0·4	5	124	82
	18·4	207	1,615	569
Requirement	18·8	207	1,615	569

The protein in the hay, silage and barley was assumed to have a degradability of 80 per cent, and the flaked maize and soya bean meal, 60 per cent (Table 2.6). The ME concentration was 11·0 MJ per kg DM with the correct amounts of both RDP and UDP being supplied in the ration.

Feeding

This topic is also described in Chapters Five, Six and Eight, and it is stressed that successful cow feeding is not a series of short-term

Table 2.6. Suggested grouping of foods based on the degradability of the protein in the rumen

Group	Degradability % (Mean and range)	Forages	Cereals	Protein supplements
A	80 (71–90)	Grass hay Legume hay Grass silage Dried grass (chopped) Swedes	Barley Wheat	Groundnut meal Sunflower meal Soya bean meal (unheated) Rapeseed meal Field bean meal
B	60 (51–70)	Grass Legumes Dried grass (ground and pelleted) Maize silage Clover silage	Maize	Soya bean meal (cooked) Coconut meal Fish meal (white) Ground peas Linseed cake
C	40 (31–50)	Dried legume (ground and pelleted)	Milo	Fish meal (Peruvian)
D	Under 31 (0–30)	Grass silage (formaldehyde-treated) Dried sanfoin (chopped)		

(Adapted from *The nutrient requirements of ruminant livestock*)

stages in which the supply of nutrients balances the exact requirements for day-to-day milk yields. Feeding in late lactation is a typical example when the cow may, in theory, be 'over fed' but body tissue is being laid down efficiently for use in the next lactation. Such terms as 'allowances' (Tables 2.3 and 2.4) and 'requirements' can give an impression of exactness and of rigidity in feeding cows, but this is misleading. There must be a dynamic approach to cow feeding. Allowances are a useful and necessary guide in planning dairy rations, but changes in the supply of nutrients can have both short- and long-term effects on milk yield. In addition food is not used solely for milk production, but for the maintenance of the body, feeding activity, growth, the development of the foetus and the production of fat and other solids in the milk. Also, as stressed earlier, the voluntary intake of dry matter varies widely at different stages of the lactation (Fig. 2.3), and is at its lowest when it is desirable to have a high intake of nutrients in early lactation.

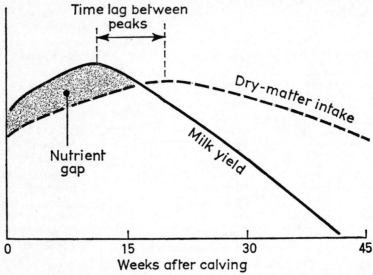

Fig. 2.4. Time lag between lactation curve and dry-matter intake of a dairy animal, and resulting nutrient gap

The major characteristics of milk yield and food intake within a lactation are shown in Fig. 2.4. The shape of a lactation curve is highly predictable, and the peak yield, often a plateau, is usually about 4 to 8 weeks after calving, whereas the maximum voluntary intake of food is reached about 12 to 16 weeks after calving. After the peak yield, the daily production of milk gradually declines at about 2 to $2\frac{1}{2}$ per cent per week until milking is stopped for the dry period before the next lactation. The shape of the lactation curve also varies according to the month of calving. The lag between peak milk yield and peak feed intake is greater in the first than in later lactations.

In early lactation, i.e. the first 10 to 15 weeks, food intake is relatively low and body reserves are used to support milk production. In mid-lactation, i.e. weeks 16 to 30, as milk yield declines, the proportion of the nutrients required to support milk production also falls, and a higher proportion of them will be diverted to replenishing body reserves. In the final 15 weeks of lactation, while voluntary food intake is still relatively high, milk yield is low and unresponsive to extra feeding.

An important factor in the lactation cycle is that body tissue is laid down with a higher efficiency in the milking compared with the non-milking cow. Dry cows do not gain weight efficiently, and their efficiency of utilization of ME is similar to that of growing cattle. It is therefore well worthwhile to use the food during the lactation period and not in the dry period, in order to lay down body tissue with maximum efficiency. These body reserves, laid down in one lactation, can then be used at an early stage in the next lactation. It is thus possible to achieve a high output of total milk, and an efficient use of feeding stuffs by making the maximum use of the cow's body reserves. Mid-lactation may be considered as the exploitation stage, with late lactation as a recovery and preparation stage of the cycle. The relationship between the price of concentrates and the value of milk will have an overall effect on concentrate use, but the approach to profitable feeding is to know and to exploit the differences in feed utilization which occur in the various stages of lactation.

The success of a whole lactation depends partly on the yield of milk achieved in the first 6 to 8 weeks, and an extra 1 kg milk per day in the peak yield will on average give an extra 200 kg milk on the final lactation yield. Generous feeding in early lactation is well worthwhile, and in this period forages of the highest quality should be given. Excessive concentrate feeding in the few days around calving may upset the cow and is not recommended.

Allocation of Concentrates

Traditionally, concentrates have been allocated on the basis of the daily yield of milk, with the highest and lowest amounts being offered in the early and late phases of lactation respectively. Thus the amount given per day could be varied each week, month, or even after a 10-week period. This system of concentrate allocation can be highly effective but requires the collection of individual milk yields and calculation of individual concentrate requirements. In recent years, simpler systems of concentrate allocation have evolved which do not treat cows as individuals. Thus, instead of offering individual cows different amounts of concentrates per day according to milk yield, the same amount of concentrates may be given to every cow in either the herd or in a group. This is called a *flat-rate system* of concentrate feeding.

In its simplest form, a fixed weight of concentrates is given to each cow during the winter-feeding period irrespective of the yield potential of the individual animal. A variation of this can be a slightly higher rate of feeding in early lactation, and a lower rate in mid- and late lactation. The system has the great advantage of simplicity and is therefore likely to be done with accuracy. Considerable research has been conducted into flat-rate systems in the last few years and it is clear that the technique will give the same amount of milk as other more complex systems. For success, good quality forage must be available *ad libitum* to all the cows. Self-feeding and easy-feeding systems using silage with a D-value of 65 are ideal.

Although it may appear that flat-rate systems of concentrate feeding are contrary to many basic nutritional principles, most of the apparent contradictions can be explained. For example, when a flat-rate is imposed, the lactation curve has a flatter shape which reduces the demand for nutrients in early lactation. In addition, the lower levels of concentrate feeding in early lactation allows the cow to consume increased amounts of silage. The deficit in nutrients in early lactation is not as great as may be anticipated and any body reserves mobilized can be replenished in late lactation.

The exact rate of concentrate feeding per cow will vary with the date of calving, the quality of the silage, and the weight of the cow, but flat-rates of 6 to 8 kg per day have been highly satisfactory.

Minerals

The mineral requirements of the cow must be adequately met if the animal is to remain healthy and capable of high milk production. Feeding trials are not particularly useful in establishing the exact amount of minerals to feed, and provide a guideline rather than an exact recommendation. The major essential elements include calcium, phosphorus, magnesium, sodium, potassium, chlorine and sulphur. Trace elements include iron, iodine, copper, cobalt, manganese and zinc. Although many minerals are termed essential to the animal when fed correctly, the same mineral can be toxic and even lethal when fed in excess. Thus the supplementation of a ration with minerals, particularly trace elements, should be done with great care.

The formulation of minerals is a specialist operation, and in certain areas it is vital to obtain local expert advice before feeding certain trace elements.

The major mineral requirements for dairy cows are given in Table 2.7. In practice, the mineral content of all feeds in the ration should be known before feeding a specific supplement. The mineral supplement can then be mixed by using limestone, salt and di-calcium phosphate, although purpose-designed commercial mixtures are generally preferable. For example, a ration containing silage, kale and sugar beet by-products can be high in Ca but low in P, and thus a mineral supplement which is high in P and low in Ca is needed. Barley and oil-seed residues such as soya bean meal are higher in P than in Ca and help to balance the minerals in the total ration. Minerals are added to the concentrate part of the diet at rates of 2 to 3 per cent of the total weight. Purchased dairy concentrates normally contain added minerals, whereas home-mixed rations must have the minerals added.

Table 2.7 Daily mineral requirements of dairy cows

	Ca	P	Mg	Na
Liveweight (kg)	*Maintenance (g/day)*			
400	14	19	6	7
500	18	26	8	9
600	21	32	9	10
	Production (g/10 kg milk)			
Fat content of milk (%)				
4·0	28	17	6	6
5·0	30	17	6	6
	Last 2 months of pregnancy (g/day)			
	17	9	2	2

(From *Animal Nutrition*)

It is possible to offer minerals *ad libitum* to cows when they are loose-housed, but the system does not guarantee that each cow receives the correct daily intake. Minerals should be given either in the concentrates or in a trough scattered on one of the feeds such as the silage. The dusting of pastures with calcined magnesite, i.e. magnesium oxide, immediately prior to grazing is one method of increasing the magnesium intake of cows and thus reducing the chances of grass tetany (Chapter Seventeen). Mineral licks have a limited use, and supply only a small weight of minerals to the cattle.

Vitamins

Vitamins are required by animals in extremely small amounts compared with other nutrients. In adult cows the two important vitamins are A and D. A mild deficiency of vitamin A can cause a scaly skin and rough hair, and a severe deficiency can cause infertility. If the cow is short of vitamin A, the new-born calf will be deficient also, and it is vital that the calf receives the colostrum, which is rich in both vitamin A and antibodies. Cows, however, are unlikely to have severe deficiencies if they have ample grazing in summer and high-quality hay and silage in winter, which build up reserves in the liver.

Vitamin D is supplied by irradiation from the sun in summer when cattle are grazing, and in sun-dried forage such as hay. There is a lack of detailed information about vitamin D and modern methods of housing and feeding, and it is possible that supplementation of the ration is required. Mineral supplements generally contain vitamins A and D and if the minerals are given at the recommended rate, they will provide an excellent way of supplying the correct amount of vitamins.

Relative Feed Costs

Although home-grown forages form the basis of most dairy rations, there are occasions when it is necessary to purchase foods other than concentrates. If this is done, it is important to assess the value of possible alternative foods so that the most economic selection is made. This can be done in an approximate way by calculating the cost of 1 t (tonne) ME and protein in the various foods, and then selecting the 'best buy'. For example it is preferable to purchase barley with an ME of 12·9 at £120 per t rather than poor hay with an ME of 7·0 at £70 per t when assessed on an ME basis. Factors other than cost, such as a minimum amount of crude fibre in the ration must be considered, but comparisons on a nutrient basis are generally well worthwhile. The extra cost of handling some foods and the possible losses in storage must not be overlooked when fixing comparative commercial prices. A more accurate and sophisticated method consists of calculating the value of ME and protein from two standard foods such as dairy compounds and barley, and using these basic unit values to evaluate other foods which are available. With this technique, a wide range of fodders, cereals, root crops and waste by-products can be accurately valued, and hence the most economic selection made.

Further Reading

Energy allowances and feeding systems for ruminants, Technical Bulletin No. 33, 1975, HMSO, London

Feeding strategies for dairy cows, ed. Broster, W. H., Johnson, C. L. and Tayler, J. C., 1980, Agricultural Research Council, London

Feeding strategy for the high yielding dairy cow, ed. Broster, W. H. and Swan, H. 1979, Granada Publishing, London

McDonald, P., Edwards, R. A. and Greenhalgh, J. F. D., *Animal nutrition*, 1981, 3rd ed., Longman Ltd, London

Nutrient allowances and composition of feedingstuffs for ruminants, ADAS Advisory Paper No. 11, 2nd edition, 1976, Ministry of Agriculture, Fisheries and Food, London

Oldham, J. D., *Protein-energy interrelationships in dairy cows*, 1984, Journal of Dairy Science

CHAPTER THREE

Grassland Production

Climate and Soil—Herbage Species—Grasses—Legumes—Seed Mixtures—Grassland Establishment—Nitrogen—Nitrogen Systems— Phosphate—Potash—Irrigation—Digestibility—Grassland Instability

---

The importance of grassland in providing food for cattle in the UK cannot be overstressed. Grass, including permanent pasture and temporary leys, occupies about two-thirds of the agricultural land, and in the major dairying areas grass is invariably the dominant crop. For example, in Cheshire and Ayrshire grass occupies 78 and 84 per cent of the total agricultural area respectively. Enclosed grassland currently contributes 67 per cent of the energy and 60 per cent of the protein consumed by ruminants in the UK (Table 3.1). Grass is a crop in its own right, which should be considered and assessed in much the same way as other crops such as roots and cereals.

Table 3.1. Contribution of grass and other sources of food to ruminants in the UK

	(*Percentage of total supply*)	
	ME	DCP
Enclosed grassland	67	60
Rough grazing	8	10
Bulk feeds	5	10
Concentrates	20	20

(From Wilkins, R. J., 'Limitations to grass for year-round ruminant production', *Forage crops—a complement to grassland*, Report of Winter 1976 meeting of British Grassland Society, Hurley, Maidenhead)

The value of grassland, and its large potential as a food, are often not appreciated because of the difficulty of measuring the yield of grassland with accuracy, especially when it is grazed. Unlike other crops, grass is not harvested directly at one specific time, and therefore it is not easy to visualize its total contribution in an extended growing season. The difficulty in measuring the yield of grass on a commercial farm is a factor contributing to the low yield which is so frequently achieved. It is vital to realize that the yield of grassland is as variable as that of other crops. Some of the factors which influence the yield of grass are not under the control of the farmer, but many others are.

Climate and Soil

The productivity of grassland is influenced fundamentally by the climate. Rainfall, temperature and sunshine affect not only the total output of herbage dry matter per year, but also the length of the growing season. Herbage growth in early spring is dominated by the temperature of the soil, and there is little growth of grass until soil temperature exceeds 4 to 5 °C. Thus, herbage growth is retarded in a cold spring, and management procedures have little effect on the date of initial herbage growth. The growth of clover starts at soil temperatures of 9 to 10°C and hence the active period of this plant is even shorter than that of grass. Grass and clover growth does not increase in proportion to rises in soil temperature, and high temperatures can reduce grass growth in the absence of water.

Rainfall may appear to be the controlling item determining the output from grassland, but the total water balance of the soil is the important factor. Evaporation and transpiration remove large amounts of moisture from the soil, and, unless this moisture is replenished by either rainfall or irrigation, a soil-moisture deficit develops and the growth of the herbage can be retarded. Moisture deficits occur in large areas of South and East England in most years, and there are deficits sufficient to limit growth even in such humid areas as South-West Scotland.

The output of grassland is also a reflection of soil type and previous soil management. When grass is grown on land with a history of arable cropping, the soil has a low available nitrogen content and yields of herbage will be low unless additional nitrogen is applied. In contrast, the soil in a grassland area will usually be rich in available nitrogen and can make a contribution to the requirements of the grass. This difference also contributes to the lower yields in the East, and the higher yields in the West of the country. Light soils which warm up quickly in spring and which do not suffer from a large soil-moisture deficit are ideal for grass growth, whereas heavy, wet soils are slow to warm up in spring and may show poorer responses to fertilizer treatments.

Herbage Species

A wide range of grass, legume and broad-leaf weed species are found in the lowland grassland of Britain, but only a few are of major importance in milk production. The main grasses are perennial ryegrass (*Lolium perenne*), Italian ryegrass (*Lolium italicum*), cocksfoot (*Dactylis glomerata*), timothy (*Phleum pratense*) and meadow fescue (*Festuca pratensis*); the legumes include white clover (*Trifolium repens*), red clover (*Trifolium pratense*), and lucerne (*Medicago sativa*). These grasses and legumes are the main species sown at the present time, but many other species invade a sown sward, especially where management is poor. These species include annual and rough-stalked meadow grass (*Poa annua* and *trivialis*), Yorkshire fog (*Holcus lanatus*) and bent grasses (*Agrostis* spp.) which are all of low productivity.

Grassland, excluding rough grazing, can be divided broadly into permanent grassland and temporary grassland, termed rotation grass or leys. Permanent grassland occupies approximately 75 per cent of the total grassland area. It is widely assumed that the output from temporary grass is higher than that from permanent pasture, but this is not generally true. Many swards sown 20 to 30 years ago can still produce high yields of grass and milk if correctly managed. Swards should not be reseeded simply because of their age, but only when their production has fallen below an acceptable level. Sward deterioration, and hence low production, result from insufficient and unbalanced fertilizer applications and the lack of a clear-cut policy of grazing, cutting and resting. Sward damage, especially in wet weather, will reduce the proportion of productive grasses and allow the ingress of poorer species.

Grasses

Without any doubt, the most useful grass is *perennial ryegrass*, which accounts for over 80 per cent of all seed sown at present. Perennial ryegrass can be used for both grazing and conservation and is one of the most persistent grasses when management is good and fertility is high. It is thus ideal either for short leys or for establishing a good permanent sward. The different varieties of perennial ryegrass are classed as early, medium and late, which is a simple classification based on the dates when growth starts in the spring and when seed heads appear. The range of heading dates from the earliest to the latest perennial ryegrass variety can span a period of about 4 weeks, and hence it is possible to select varieties which will produce a sequence of leafy herbage in the spring for both grazing and con-servation.

Recommended varieties of grasses are listed by the National Institute of Agricultural Botany and the Scottish Agricultural Colleges after extensive testing. Recommendations are slowly chang-ing as new varieties are tested and older ones discarded, but two of the oldest varieties, S24 and S23, are still grown.

Italian ryegrass is a vigorous and high-yielding grass which starts growth in the spring before perennial ryegrass. Most varieties persist for only 18 to 24 months after sowing, and hence it is a grass for short leys of either 1 or 2 years' duration. Italian ryegrass can produce an 'early bite' for dairy cows (Chapter Four), followed by either a leafy hay or silage crop and then late grazing. This grass does not produce a dense, tight sward, and because of its 'open' nature, it will poach badly in wet weather, i.e. the sward will become muddy and the plants be damaged during grazing. Italian ryegrass is a grass with specialized uses, and although it can fit well into certain systems of grass production for milk it is not as widely grown as perennial ryegrasses, although its use for conservation has increased in recent years. A recommended variety of Italian ryegrass for most situations is RvP.

Alternatives to Italian ryegrass are the *hybrid ryegrass* varieties such as Augusta, but many of these grasses are susceptible to killing in the winter in frosty districts and cannot be grown widely. Augusta is a tetraploid grass, i.e. a variety containing twice the number of chromosomes found in the normal diploid varieties. Tetraploids have larger cells than diploids and thus have a lower content of dry matter, thicker stems, and more sugar. Tetraploids are highly pala-table and are selected by cattle in preference to diploids, but are more difficult to conserve as hay. The high content of sugar in tetraploids is an advantage in silage making (Chapter Six), although their lower dry-matter content makes some wilting advisable.

Timothy, *cocksfoot* and *meadow fescue* are all much less important than ryegrass. Timothy persists in cold, wet conditions and will give heavy crops of late hay of low digestibility. It is best sown either alone or with meadow fescue, which is not a vigorous species. Cocksfoot has a low palatability and low digestibility and may become dominant in swards which are undergrazed or mown for hay year after year. Its use in milk production is virtually nil. A more useful and palatable grass is meadow fescue, but it is difficult to establish.

Legumes

The most important legume is undoubtedly *white clover*, which can make a valuable contribution to the supply of nitrogen to the grasses in the sward. In addition this plant thrives under hard grazing conditions, is highly palatable, and maintains its high digestibility throughout the growing season. It is recommended for sowing in all seed mixtures (Table 3.2) which will last for 3 or more years, even where the use of heavy dressings of fertilizer nitrogen may eventually eliminate the clover.

White clover is classified on the basis of leaf size. The small-leafed varieties, such as Kent Wild White, persist under grazing conditions, are hardy and are included in leys of 3 or more years. The medium-leafed varieties, including the famous S100, make a larger contribution to the sward under more lenient grazing conditions and are invaluable in medium-duration leys. The large-leafed white clovers have generally proved to be poorly adapted to British conditions, but the variety Blanca RvP appears to combine persistency and winter-hardiness with the ability to counteract the competition of the grass when heavy dressings of nitrogen are applied.

Red clover is normally a short-lived species for use in 1- and 2-year leys (Table 3.2). This legume can produce high yields of dry matter without the need for fertilizer nitrogen and can be a useful conservation crop for silage, although it has a low content of dry matter when cut (Chapter Seven). The tetraploid varieties Redhead and Astra, which are classed as early and late respectively, are high yielding and excellent for silage making.

Lucerne is a crop for the drier and warmer parts of Britain, and can produce high yields for conservation without the need for fertilizer nitrogen. Lucerne has a long growing season, and in the South of England three or four crops are possible in the season, given good management. Lucerne does not thrive in wet, acid, and

Table 3.2. Simple seed mixtures for short and long leys

	Maximum values (kg per ha)	
1-year ley	*Mixture 1*	*Mixture 2*
Italian ryegrass	35	14
Perennial ryegrass	—	14
Red clover	—	4
Total	35	32
2-year ley	*Mixture 3*	*Mixture 4*
Italian ryegrass	10	—
Perennial ryegrass	20	8
Red clover	—	14
White clover	2	—
Total	32	22
Long leys	*Mixture 5*	*Mixture 6*
Perennial ryegrass, early	26	—
Perennial ryegrass, late	—	26
White clover	2	2
Total	28	28

(Adapted from *Seed mixtures* and *Seed mixtures for Scotland*)

waterlogged conditions. The variety Europe is recommended for general use, and Vertus in areas where verticillium wilt is serious. Lucerne is primarily a crop for conservation, but in some limited circumstances it can be either grazed or cut and fed green to cattle.

Seed Mixtures

When grassland is resown it is important to select a seed mixture which has been designed specifically for the intended purpose of the ley. In the past a multi-purpose mixture containing six or more species was common, but now the preference is for special-purpose simple mixtures with only one or two species. Some typical simple mixtures are given in Table 3.2.

The rates of sowing given in Table 3.2 are maximum values, but where soil and seed-bed conditions are favourable the rates can be reduced by 5 to 6 kg per ha. If the seed is drilled in good conditions a further reduction of about 4 kg per ha is possible. Rates of sowing are generally higher in Scotland than in Southern England, and less seed can be used when undersowing a cereal crop than for a direct reseed. Mixture 1 requires heavy dressings of fertilizer nitrogen whereas mixtures 2, 3 and 4 can receive either low rates or none. Mixtures 5 and 6 will both respond well to high levels of fertilizer nitrogen for grazing and for silage. Either single varieties or mixtures of varieties within a species, i.e. early or late, may be used within a mixture. It is emphasized that the selection of a seed mixture is only one step in growing grass and producing milk, and detailed attention must be paid to many other important factors.

Grassland Establishment

High priority must be given to the correct establishment of the grass sward, which can be sown either direct or under another crop. Direct reseeding is the sowing or drilling of the grass and clover seeds, without any other crop, into a carefully prepared seed bed. When a fine, firm and moist seed bed is available in the spring, the seeds germinate rapidly and a first-class ley can be established in the year of sowing. Seeds should be sown early, e.g. in April, and should normally be grazed in June for the first time, and the sward topped with a mower, with a repeat of this procedure every 3 to 4 weeks. If the seeds cannot be sown in early April, sowing should be deferred until August or even early September, depending on the district and the rainfall. The later the seeds are sown in autumn, the greater is the risk that clovers and less persistent grasses will not establish themselves successfully. Total grass production from a ley in the sowing year will be about half that from an established ley, but the new grass is available at a vital time for the dairy herd when other grass is scarce.

An alternative method of establishment is to sow the grass seeds under a cereal crop which is harvested in the autumn. Grass is not available until after harvest time and if the cereal crop is flattened by the weather the grass can be killed. Where grass is an important and integral part of the farming system, it is preferable to sow the seeds direct and not under cereals. A compromise solution is to sow the grass seeds with either cereals which are cut early for silage or Italian ryegrass which is grazed down quickly. The grazing of a ley in the year of sowing requires care; the preferred method is extensive grazing with no concentration of stock on one small area, and the avoidance of grazing on wet days when sward damage can occur.

A newly established ley will normally contain some annual weeds, which can be controlled by cutting. Spraying with a weedkiller should only be required on rare occasions, and it is then vital to select the correct chemical formulation if the clover in the sward is not to be killed. The spraying of established pastures to reduce broad-leafed weeds, e.g. docks, and some less productive grasses is possible, but the cost of the operation must be weighed against the possible increase in herbage yield and animal production.

For the successful establishment of grass, and also, to a greater degree, of clover, the soil must not be acid: a pH of 6 is about ideal. To obtain the correct pH, ground limestone or magnesium limestone should be applied. An adequate amount of phosphate is needed to ensure that the roots of the grass and clover seedlings are well established, and a seedbed application of 60 to 70 kg P_2O_5 per ha is strongly recommended. Potash and a small amount of nitrogen are also beneficial at this stage in the development of the ley and 30 kg nitrogen and 50 kg K_2O per ha are suggested. These fertilizer nutrients should be applied prior to the sowing of the seed and must be harrowed well into the seed bed.

A good 'take' of seeds is the foundation on which future high yields of grass are built, and every effort must be made to establish a vigorous, weed-free sward in the first year. Swards intended for long leys should preferably be grazed in the year of sowing and in the first harvest year, i.e. the year after sowing, as cutting for conservation will reduce tillering (side-shoot formation) and produce an open sward lacking in clover. Short leys containing Italian ryegrass can be cut and grazed as required.

Nitrogen

The yield of grassland is extremely sensitive to the supply of nutrients, especially nitrogen (Fig. 3.1). In this example, the sward of perennial ryegrass (S23) was cut five times in the growing season when the grass was 150 to 220 mm high. Clearly, when no nitrogen fertilizer was applied, the yield of dry matter was extremely low, but as increasing amounts of nitrogen were applied up to 350 kg per ha, the yield of herbage dry matter increased with a response of approximately

30 kg dry matter per kg of nitrogen applied. At rates above 350 kg per ha, the response became much smaller and was negligible at the highest rates of application. Within broad limits, dry-matter production is directly proportional to the quantity of nitrogen applied and, at the present price of nitrogen, an application rate of 350 to 400 kg nitrogen per ha is economic if the herbage is utilized efficiently for milk production.

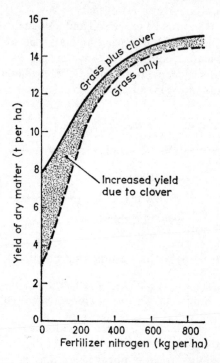

Fig. 3.1. Annual yields of dry matter from grass and grass/white clover swards receiving a wide range of fertilizer nitrogen rates per ha (From D. Reid, The Hannah Research Institute, Ayr, 1977)

Fertilizer is not the only source of nitrogen for the sward. An average of 90 kg nitrogen per ha may be supplied from the soil, which is equivalent to an application of about 130 kg fertilizer nitrogen per ha. Clover also supplies nitrogen, and in a well balanced grass/clover sward about 140 kg nitrogen per ha is produced. This amount varies widely, depending on the clover content of the sward, but in general it is well worth having even when fertilizer nitrogen is applied.

The effect on yield of including clover in a grass sward at different levels of fertilizer nitrogen application is also illustrated in Fig. 3.1. At low levels of fertilizer input, the value of the legume in the sward is clear, although in commercial practice it is not easy to maintain clover in a sward receiving higher levels of fertilizer nitrogen. Clover has a much shorter growing season than grass, and on an intensive dairy farm with a high rate of stocking it is impossible to rely on clover as the only source of nitrogen.

Nitrogen is also supplied to the soil and ultimately to the grass via the dung and urine of the grazing animals, and at high levels of herbage production, e.g. 12 t dry matter per ha, about 120 kg nitrogen per ha may come from this source. A diagrammatic representation of the nitrogen supply to the plant is given in Fig. 3.2.

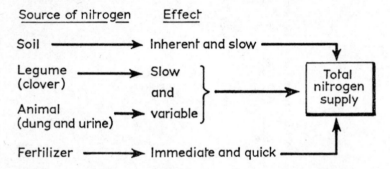

Fig. 3.2. Main sources of soil nitrogen which are available to grass (Based on J. S. Brockman, *The role of nitrogen in grassland productivity*)

There are many interactions between these four main sources of soil nitrogen, but the all-important fact is that for high production from grassland a high total input of nitrogen is required.

Nitrogen Systems

From this information on sources of nitrogen (Fig. 3.2) it is possible to plan various systems of nitrogen management on the dairy farm according to the desired intensity of the stocking rate. If the overall aim is to produce high yields of grass dry matter, e.g. 11 to 12 t per ha in the season, there is no alternative to an application of 300 to 360 kg nitrogen per ha split into five or six separate dressings. With this input of nitrogen, clover will only play a small part in the total supply of nitrogen, but soil and animal nitrogen will each make useful contributions. Each grazing will require an application of about 60 kg nitrogen per ha, and for conservation cuts the rate must be increased to 80 or 100 kg nitrogen per ha (Table 3.3). Nitrogen fertilizer should be applied as soon as a field is vacated after grazing or cutting, since a delay at this time will cause a serious loss of yield.

Table 3.3. Suggested application rates of fertilizer nutrients for grassland

	(kg per ha)		
	N	P_2O_5	K_2O
Grazing, intensive			
Spring	60	—	—
After 1st grazing	60	20	20
After 2nd grazing	60	20	20
After 3rd grazing	60	20	20
After 4th grazing	60	—	—
After 5th grazing	60	—	—
Total	360	60	60
Grazing, extensive, clover used			
Spring	60	—	—
Summer	—	20	20
Autumn	60	20	20
Total	120	40	40
Silage, intensive			
1st cut	100	30	80
2nd cut	90	20	70
3rd cut	80	20	40
Autumn grazing	60	—	—
Total	330	70	190

In marked contrast to this high-fertilizer system, it is possible to rely almost entirely on clover to supply the nitrogen and to use little or no fertilizer nitrogen. Some nitrogen will come from the soil and the stock. Herbage production will be 6 to 7 t dry matter per ha, the grazing season will be short and the rate of stocking will be about half that on the high-fertilizer system. In a few highly favoured areas a grass/clover system is possible, but in most areas with high land values and high fixed costs the system is not viable.

In a third system, use can be made of both clover and fertilizer nitrogen by a technique which is midway between the two previous systems and which will produce about 8 to 9 t dry matter per ha. Most of the swards will contain clover, which will make a contribution to mid-season production. However, nitrogen will be used to produce early grass in spring and to grow herbage for conservation. The fertilizer nitrogen and the clover will not be antagonistic, and each will contribute to producing herbage efficiently and economically. On many farms this system, with an average use of about 125 kg nitrogen per ha, could be used with success to extend the grazing season and produce more conservation crops.

Table 3.4. Grass production and milk output from four systems of sward management with clover and fertilizer nitrogen

Type of sward	Fertilizer nitrogen (kg per ha)	Dry-matter yield (kg per ha)	Cows per ha	Milk (kg per ha)
Grass	Nil	3,500	1·25	3,000
Grass + clover	Nil	7,000	2·50	6,000
Grass + clover	125	8,500	3·00	8,000
Grass	360	12,000	4·00	11,000

A general indication of the output from some alternative systems of nitrogen use is given in Table 3.4. In this Table, each kg of herbage dry matter produces approximately 1 kg milk, but it is stressed that an increase in the yield of grass does not always imply an increase in either the yield of milk or the gross margin per ha. Only when the grassland is efficiently utilized at the correct stocking rate will the response to nitrogen in terms of milk yield per ha be similar to the response of dry matter indicated in Fig. 3.1.

Phosphate

In order to obtain the maximum response from nitrogen, it is necessary to ensure that other fertilizer nutrients such as phosphate and potash are not limiting production. Established grassland does not normally require a high input of phosphate: an average annual application of about 40 to 50 kg P_2O_5 per ha is adequate. Where heavy crops of silage are being cut and removed, more phosphate is needed than under grazing conditions, but in general phosphate is applied to maintain the reserves in the soil. A few extremely phosphate-deficient soils will respond markedly to applications of phosphate fertilizers, and responses are most obvious in cold, wet springs. A crop of 10 t herbage dry matter per ha will contain 60 to 80 kg P_2O_5, but some of this is recycled, somewhat inefficiently, via the animal. Phosphate may be applied either once per year or as a number of split dressings (Table 3.3).

Potash

Potash is essential for herbage growth, and a shortage of this nutrient will rapidly lower the response to nitrogen. A lack of potash will also encourage poor, unproductive grass species into the sward, and the more productive grasses will disappear. Unlike phosphate reserves in the soil, potash levels can be depleted quickly where high rates of fertilizer nitrogen are applied. A herbage yield of 10 t dry matter per ha will remove 200 to 300 kg K_2O from the soil, and this must be replenished if high yields of herbage are to be maintained, but there is a vast difference between the use of potash on grazed swards and on cut swards. On grazed swards high potash levels in the herbage are undesirable, since this reduces the magnesium level in the herbage and may cause hypomagnesaemia in the cows. Potash is also returned to the soil via the cows' urine.

In addition, grass can absorb more potash than it requires for growth, and this 'luxury' uptake is both wasteful and dangerous. Thus under intensive grazing conditions with an input of 360 kg nitrogen per ha, only 50 to 60 kg K_2O per ha is needed. This potash should not be applied before the first grazing in spring, and it is advantageous to divide the total amount between two or three separate dressings in mid-season. If slurry has been applied to a grazing sward—and this practice is not recommended—no potash fertilizer should be applied. Skill and experience on the individual farm are required to determine the exact level of potash needed under grazing conditions.

On swards which are to be cut, potash is essential for high production, and 100 kg nitrogen will require balancing with about 80 kg K_2O per ha. Potash is required for each separate cut of hay and silage, but if the aftermath is to be grazed, no more potash should be applied. Slurry is high in potash and is ideal for fields which are to be cut and conserved. It is most unwise and uneconomic to apply one large application of potash for the entire growing season.

A general summary of the suggested recommendations for fertilizing grassland is given in Table 3.3. Fertilizers containing only one nutrient may be used, but compound fertilizers which have been specially formulated are preferable if they are selected with care. Slurry is a valuable source of fertilizer nutrients, and is discussed fully in Chapter Thirteen.

Irrigation

Although irrigation is normally associated with arid and semi-arid regions, additional water will increase herbage production in most parts of lowland Britain in one year out of two. Typical responses range from 100 to 300 kg dry matter per ha for every 10 mm of irrigation water applied.

Successful irrigation of grassland makes a number of demands. An adequate source of water is of course essential, and this should be available throughout the summer growing season. The most convenient source of water is a river or stream which flows through the farm, but permission must be sought before using it. Boreholes and small dams are possible, but are much more expensive and also require permission. It is also important to have enough reliable equipment to maintain the soil moisture deficit at a level of 25 to 50 mm. Finally, the grassland to be irrigated should contain a high proportion of productive grasses and be adequately supplied with nitrogen. The exact timing and the amount of water required can only be calculated locally, and this should be done in conjunction with either the meteorological office or the agricultural advisory service.

The direct value of irrigating grassland is not easy to calculate, and is made more complex if irrigation allows more fertilizer nitrogen to be applied. Clearly, if irrigation increases the yield and quality of grassland, more cattle can be kept on the same area or some grassland can be released for another crop. A further advantage is that a continuous supply of high-quality leafy grass will help to maintain milk yields during a period of the year when yields tend to fall, and thus concentrates can be saved.

Grassland irrigation is really an insurance premium against summer drought, and the annual cost of water, equipment and labour has to be balanced against an increase in herbage yields and herbage quality, and hence more milk. It is stressed that irrigation is a precision job involving high capital expenditure, which is only justifiable where the standard of grassland management is high.

Digestibility

As grasses and legumes grow and mature during spring and early summer the yield of dry matter per ha increases (Fig. 3.3) but the digestibility, expressed as D-value, falls. At a specific date different grasses can differ markedly in D-value (Table 3.5), and this is not

Table 3.5. D-values of grasses at different dates

May	Cocksfoot, early	Perennial ryegrass, early	Italian ryegrass	Perennial ryegrass, medium	Perennial ryegrass, late
5–9	—	70	—	—	—
10–14	67	—	70	—	—
15–19	*	*	—	70	—
20–25	63	67	—	—	70
26–31	—	—	67*	—	—
June					
1–5	—	63	—	67*	—
6–10	—	—	63	—	67*
11–15	—	—	—	63	—
16–20	—	—	—	—	63

*Dates when 50 per cent of flower heads have emerged
(Based on *Grasses and legumes for conservation*)

related exactly to the date when 50 per cent of the flower heads have emerged. The D-value of white clover is higher than that of red clover and lucerne (Fig. 3.4) and of many grasses. Consequently, grass/clover swards have higher D-values than pure grass swards.

If the herbage is cut, the regrowths may not have as high a D-value as the first spring growth and a decline in value with time will occur as before. As the number of cuts in the season increases, the average D-value of all the herbage will increase also, although the output of digestible nutrients per ha will fall. To overcome this reduction in output per ha, the amount of nitrogen applied per season should be increased. If this is done, the conflict between output per ha and D-

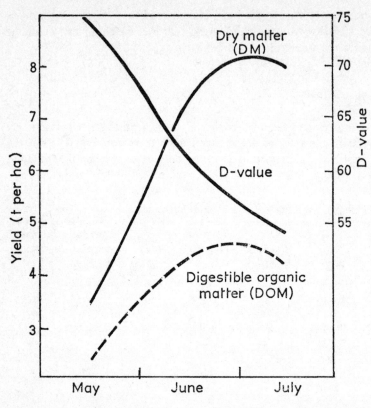

Fig. 3.3. D-value, yield of dry matter and digestible organic matter of an early perennial ryegrass (S24) during the first growth
(From R. D. Harkess, West of Scotland Agricultural College, 1977)

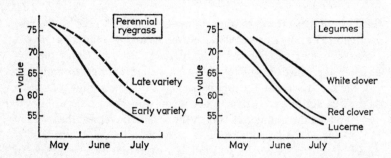

Fig. 3.4. D-values of two varieties of perennial ryegrass and three legumes
(From R. D. Harkess, West of Scotland Agricultural College, 1977)

value is reduced markedly. If herbage is adequately fertilized and harvested approximately every 4 to 5 weeks, the output of digestible nutrients per ha is maximized.

In a grazing system, where the grass is eaten and then rested for 3 to 4 weeks, the herbage remains leafy with a D-value of 68 to 72. This is a high value and, because of the highly selective nature of the grazing cow, the herbage eaten will have a D-value of about 80. This value is similar to that of a concentrate, and is one reason for the high daily milk yields which are possible from high-quality leafy pasture.

For conservation as hay and silage it is usual to cut herbage at a more mature stage than the grazing stage and hence the D-value will be lower. A D-value of 67 has been adopted as the main standard for comparing species and varieties of herbage (Table 3.5), although this value will not be correct for all situations. D-values for conservation crops are discussed fully in Chapters Five and Six, but in general there is a need to increase the D-value of both hay and silage by earlier cutting. It is emphasized that although grass may be cut at a known level of D-value, the ultimate value of the conserved crop may be different because of the inefficiency of the conservation process.

An important point to note in Table 3.5 is that different grasses reach the same D-value, e.g. 67, over a wide span of dates. This is of considerable significance when organizing grassland conservation systems. If several fields on the farm are each sown to different varieties of ryegrass, a succession of first growths of similar digestibility can be obtained over a period of 3 to 4 weeks. If labour and

Table 3.6. Method for estimating the D-value of herbage from the stage of growth and the proportion of leaf

		No flower heads emerged	Flower heads just emerging	Flower heads 75% emerged	Heads emerged and free from top leaf	Heads emerged and flower stalks elongating	Heads emerged and anthers visible
Weight of leaf as a percentage of total crop	10	—	—	—	—	57	55
	20	—	—	63	61	60	58
	30	69	67	66	64	63	61
	40	72	70	69	67	—	—
	50	75	73	—	—	—	—

(Adapted from R. J. K. Walters, 'The field assessment of digestibility of grass for conservation', ADAS *Quarterly Review*, No. 23, 1976, p. 323)

machinery are limited, silage making can be extended over several weeks and the crop will have a fairly uniform digestibility.

The results in Table 3.5 apply generally to Central and Southern England and Wales, but the dates are only a guide and vary slightly in early and late seasons. The dates tend to be later in the North, but there are exceptions such as South-West Scotland, which is an early district.

A rapid field method for assessing the D-value of grass, based on a visual appraisal of the stage of growth and the percentage of leaf, is given in Table 3.6.

Grassland Instability

Grassland in Britain is an extremely unstable type of vegetation. Grass, and the associated plants in a pasture, are only one early stage in the complex changes in vegetation that would eventually replace the grass sward by scrub and then by woodland. The sown ley is even more unstable than permanent pasture, and it must always be realized that changes in pasture management will alter the botanical composition of the sward and its level of production. Drainage, liming, fertilizer application and grazing are all potent factors which influence the type of sward and its output. Grassland production is conducted in an artificial, man-made environment, and this instability should provide a challenge and thus add further interest to the task of economically producing milk from grass.

Further Reading

Camlin, M. S. and Gilliland, T. J., 'Grasses—recommended varieties 1982–83', *Agriculture in Northern Ireland*, 57, 1982, p. 2.

Classification of grass and clover varieties for Scotland, 1983–84, Publication No. 90, 1983, Scottish Agricultural Colleges, Auchincruive, Ayr

Holmes, W. (ed.), *Grass, its production and utilization*, 1980, The British Grassland Society, Blackwell Scientific Publications, Oxford

Jollans, J. L. (ed.), *Grassland in the British economy*, Paper No. 10, Centre for Agricultural Strategy, 1981, Reading, Berkshire

Recommended varieties of grasses, Farmers' Leaflet No. 16, 1982–83, NIAB, Cambridge

Recommended varieties of herbage legumes, Farmers' Leaflet No. 4, 1982–83, NIAB, Cambridge

Seed mixtures 1983, Booklet 2041, 1982, Ministry of Agriculture, Fisheries and Food, London

Seed mixtures for Scotland, Publication No. 86, 1982, Scottish Agricultural Colleges, Auchincruive, Ayr

Thomas, C. and Young, J. W. O. (eds.), *Milk from grass*, 1982, I.C.I. Agricultural Division, Billingham and Grassland Research Institute, Maidenhead

CHAPTER FOUR

Grazing Systems

*Continuous Grazing—Stocking Rate for Continuous Grazing—
Rotational Grazing—Paddock Grazing—Rigid Rotational Grazing
—Strip Grazing—Leader-and-follower System—Types of Fencing—
Water Supply—Extending the Grazing Season—Autumn Grazing—
Supplementary Feeding—Stocking Rate—Mechanical Treatment—
Zero Grazing—Target Milk Outputs*

Grassland can produce high yields of dry matter and digestible nutrients per ha if it is treated as a crop and managed correctly. Unfortunately, the utilization of the grass, i.e. the conversion of the herbage to milk, is often extremely poor. In order to produce milk from grass it is vital to grow sufficient grass, but it is equally necessary to convert that herbage into milk by efficient systems of grazing. Grass is only a means to an end, and on dairy farms the object is to sell milk.

The potential yield of milk from grass is high. A Friesian cow weighing 600 kg and with an intake of 3 kg dry matter per 100 kg liveweight can eat 18 kg of herbage dry matter per day, which can supply 218 MJ. This amount of energy is sufficient to maintain the cow, give a liveweight gain of 0·5 kg per day and produce 28 kg milk per day. Individual milk yields of this level are not unknown from grass alone, but in commercial practice it is often difficult to reach and to maintain this value.

Under experimental conditions where high rates of fertilizer nitrogen were applied and stocking rate was high, the output of milk has reached almost 17,000 kg per ha. On a commercial farm, such an output is unlikely to be achieved, but the figure indicates the high potential of well-managed grassland. Yields of 11,000 kg milk per ha (Table 3.4) are certainly possible under commercial conditions.

The aim of a good grazing system is to provide the cow with a regular daily supply of highly digestible grass which will match the requirements of the animal for as long a grazing season as possible.

It should integrate the supply of nutrients from the grazing and the demands of the cow, and should ideally be a relatively simple system leading ultimately to profitable milk production on the whole farm.

The importance of having a definite system of summer grazing is still not appreciated. For example, in a recent census of over 3,000 milk producers in Scotland 21 per cent had no grazing system for their dairy herd. Set stocking was stated to be practised on 46 per cent of the farms, and the remainder used paddock grazing and strip grazing. The choice of a specific grazing system depends on the layout of the farm, the size of the fields, the number of cows, and the availability and interest of the labour. The main systems are continuous grazing, rotational grazing and strip grazing, and there are numerous variations within each of them. The importance of the type of grazing system is small however, compared with that of producing sufficient herbage and having the correct rate of stocking.

Continuous Grazing

Continuous grazing or set stocking is a simple system in which the grazing area for the entire dairy herd is in one large block with no internal subdivisions, and the animals graze over the whole area for the entire season (Fig. 4.1). Management is minimal and few decisions are required. Renewed interest in this system stems from changes in fertilizer management and higher rates of stocking.

The success of the system depends on the amount of herbage produced by the sward, the overall intensity of stocking and the flexibility of the stocking rate over the grazing season. If sufficient herbage is to be available to the cows, it is imperative that the sward receives about 2 kg fertilizer nitrogen per ha per day. Thus, over a 180-day grazing season, 330 to 360 kg nitrogen per ha would be required. This amount of fertilizer should be applied in five or six separate applications starting with an initial dressing about 3 to 4 weeks before the grazing starts, and then followed by monthly applications over the whole area. To make optimum use of the available rainfall it may be preferable to split each monthly dressing of fertilizer into two, and apply each one at intervals of 2 weeks to alternate halves of the grazing area. Further subdivision of the fertilizer for weekly applications does not appear to be worthwhile. However, a regular disciplined system of fertilizer application is desirable. The fertilizer can be applied without removing the animals, and, because of the large area, the application can be done quickly and evenly.

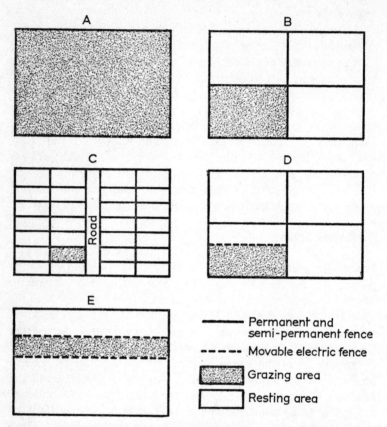

Fig. 4.1. Diagrammatic layout of five grazing systems for dairy cows *Key:* A = continuous grazing or set stocking; B = rotational grazing with 4 7-day paddocks; C = paddock grazing with 28 1-day paddocks; D = rigid rotational grazing system with 4 paddocks each grazed for 7 days with front electric fence; E = strip grazing with a front and a back electric fence

Stocking Rate for Continuous Grazing

Stocking rate, i.e. the number of cows per ha, is a vital factor in this, and indeed in all grazing systems. Should the sward be overstocked, the cows will not eat enough herbage and milk yields will fall. On the other hand, with a low rate of stocking the herbage will become mature, and although the cows may select an adequate diet and maintain milk yields, the output of milk per ha will be low. Thus for success, the stocking rate must not be fixed for the entire season, but vary in order to match the needs of the cows with the production of the sward. Over the whole grazing season a stocking rate of 5 cows per ha is desirable.

It is stressed that this is an average rate, and a herd calving in spring may be stocked fully with 4 cows per ha. Conversely, an autumn-calved herd may be stocked at a higher rate. Overall stocking rate depends on the yield of herbage, which is itself a reflection of many factors which include the soil fertility, the fertilizer input, the rainfall and sward quality. Flexibility of stocking rate within the grazing season will generally mean that the area allowed per animal will increase slowly but steadily as the season progresses. In April and May, when the grass is growing rapidly, 6 to 7 cows per ha may be needed; whereas in August and September, 2 to 3 cows per ha may be adequate. As cow numbers in a dairy herd are normally static, the variation in stocking rate is made by increasing the grazing area by bringing in aftermaths from the fields cut for conservation. As in all systems of grazing, the management of the grass must be dictated by the output of milk and the condition of the cows.

Leys containing pasture types of perennial ryegrass such as S23, and permanent pasture with productive grasses, are ideal for continuous grazing. Swards will thicken and improve with continuous grazing if stocking rate is controlled and adequate amounts of balanced fertilizer are applied. Grazing should not be delayed in the spring until too much herbage is available to the cows, and the grass should be allowed to grow to the cows.

Continuous grazing requires less fencing and labour than other systems, but the cows may be scattered over a wide area, and rounding them up at milking time can be tedious and time-consuming. Probably the biggest disadvantage of the system is that milk production per ha is often 10 to 15 per cent lower than on a rotational system of grazing at the same stocking rate. As the rate of stocking increases, the difference in output between continuous grazing and rotational grazing will increase, and at times of severe grass shortage the continuous grazing system can 'break'. Nevertheless, the system has a definite place for many milk producers. An output of over

11,000 kg milk per ha has been obtained from continuous grazing—
a highly satisfactory level of production.

Rotational Grazing

Rotational or paddock grazing is a system which allows the major
part of the grazing area to be rested and hence to grow, while a
smaller area is being grazed rapidly by the cows. As the frequency of
herbage defoliation is increased, the amount of herbage produced
decreases, but the quality of the material increases. The art of
rotational grazing is to strike the correct balance between quantity
and quality of grass and then to devise a system of grazing and
resting the grass in order to achieve this balance.

A simple system of rotational grazing is one based on 4 separately
fenced paddocks of equal size (Fig. 4.1). One paddock is grazed by
the cows for 1 week, fertilizer is then applied, and the area is rested
for 3 weeks before the next grazing. This simple rotation allows the
herbage to regrow and achieves a suitable compromise between the
yield and quality of herbage. In practice, a total grazing area is
allocated which allows 1 ha for 5 cows, and this area is then divided
into 4 paddocks of approximately the same size. Compared with
continuous grazing, more fencing and more water troughs are
required, and some land is usually lost to form corridors to the
paddocks. A major problem with only 4 paddocks is that herbage
growth is not uniform at all periods of the season, and thus the
amount of herbage on offer to the cows in different months will vary
widely. This problem can be partly overcome by either altering the
amount of fertilizer applied at each dressing or by taking one of the
4 paddocks out of the grazing rotation and cutting it for conservation. These and other possible modifications tend to make the system
more complex and involve more decisions, and can lead to two
distinct ways of modifying the rotational grazing system.

Paddock Grazing

The first method is to increase the number of paddocks from 4 to
between 21 and 30, each one being fenced separately and with its own
gate and access to water (Fig. 4.1). If there are 28 paddocks in the
system, only one is grazed each day, and thus the remaining paddocks

have 27 days in which to regrow. The overall rate of stocking is again 5 cows per ha, but with a concentration of 140 cows per ha on one paddock each day. When there is an excess of grass, a few paddocks can be removed from the rotation by cutting them for conservation, and thus the cycle of regrowth is maintained for periods when grass growth is not so rapid. The total application of fertilizer nitrogen can vary from 330 to 360 kg per ha in the season, with 50 to 60 kg per ha applied to each paddock each month. The fertilizer should be applied immediately the cows have grazed a paddock, but it is often more practicable to wait until 2 or 3 paddocks can be dressed on the same day.

The regrowth from paddocks cut for conservation is normally much slower than the regrowth from grazed paddocks, and only the minimum number of paddocks should be cut. A system of 28 1-day paddocks gives a more flexible approach to rotational grazing than one with only 4 paddocks, but skill is required to balance the needs of the cows accurately with the correct amount of herbage. The system demands a limited number of decisions, but in general it imposes a discipline on the operator, who realizes quickly the importance of providing a regular daily supply of grass to the dairy herd.

The amount of fencing required for 30 small paddocks is obviously much larger than that for 4 large paddocks, and more land can be wasted as corridors. However, with careful planning and the use of the normal field boundaries, fencing costs can be kept to a minimum. Similarly, water troughs can often be shared between 2, or even 3 separate paddocks. Milk yields of 12,000 to 15,000 kg per ha have been obtained from a system of 1-day paddocks, and with experience it can lead to a fuller understanding of the principles of grassland production and utilization. Problems can arise if the herd size increases and when machines have to enter the paddocks, but these are all relatively small drawbacks which can be overcome.

Rigid Rotational Grazing

The second method of rotational grazing is to retain the 4 original paddocks, but to allow the cows only 1/7th of one paddock each day by the use of a movable electric fence (Fig 4.1). This is a rigid combination of rotational grazing and daily strip grazing, and it builds a further degree of control into the system. This technique is referred to as the Wye College system, and incorporates a rigid system of moving from paddock to paddock each week regardless of the amount of grass available on a paddock. Cutting for conservation is not done, and thus the utilization of the herbage on a plot may be low at one grazing, but substantially higher at the next, herbage often being carried over from one grazing cycle to another. It is important to start grazing early enough in spring to ensure that the last of the 4 paddocks to be grazed is not too mature.

The production of herbage is high because of the input of 330 to 360 kg fertilizer N per ha, herbage regrowth is good because of the 21-day rest period, the cows obtain a fresh area of grass each day, and the system requires few management decisions. Over a 4-year period, the output of milk from this system has been 12,000 to 15,000 kg per ha—virtually the same as the yield from a 28-paddock system at the same rate of stocking. Compared with the 28- to 30-paddock system, there are savings in fencing, gates, water supplies and corridors. Possibly the most attractive feature of the rigid rotational system is its complete lack of daily and weekly management decisions, which gives it the great merit of simplicity and ease of operation.

All systems of rotational grazing have the problem that the day-to-day intake of nutrients from the herbage is subject to variation, with a consequent fluctuation in milk yield. In the 30-paddock system the amount of herbage on offer on successive days is rarely the same, and in the 4-paddock system the intake of herbage varies from day to day because of the grazing habits of the cows. On day 1, when entering a fresh paddock, the cows select the highly digestible tips of the grass and milk yield increases. On days 5 to 7 in the same paddock, the cows have to consume the stemmier and less digestible parts of the grass and milk yield falls. Cows should be removed from a paddock before the milk yield declines too rapidly, or alternatively the daily allocation of herbage should be increased by skilful use of a movable electric fence.

Strip Grazing

This system, which is also called close folding, is the one grazing technique which can fairly accurately supply the herbage requirements of the cows on a day-to-day basis. In essence, an electric fence is moved forward either once or twice per day to allow the cows a sufficient ration of fresh grass to maintain a normal pattern of milk yield without any violent day-to-day fluctuations. If a skilled operator is both moving the fence and closely watching the daily milk yields, the system is possibly the best one available, but its apparent simplicity is deceptive. Although variations in the area of land allowed to the herd can be made each day, it is absolutely vital to plan ahead, and to know exactly where the cows will be grazing in 1 to 2 weeks' time. Skill is demanded in forward planning, which includes the application of fertilizer, resting fields, and the bringing in of silage and hay aftermaths.

Decision making is a daily task, and in addition to a front fence (Fig. 4.1) a back fence is desirable if a field is to be grazed for more than 1 week. The daily task of fence moving can be difficult when the soil is hard, but this job should be regarded as one of the most important single jobs on the farm in the summer.

Moving the fence the correct distance is the main link between the herbage and the milk yield in the summer months. Because the cow cannot unduly select parts of the grass when strip grazing, and has to eat all the herbage on offer, the animal is almost wholly dependent on the ability of the stockman to provide suitable material in the correct amount.

To utilize the herbage efficiently, it is wise to give the herd a long, narrow strip of fresh grass at each grazing. This avoids soiling and treading of the herbage and decreases wastage. At times, much of the available herbage is below and beyond the electric fence and the cattle must reach for it. At other times, the front fence may be moved forward in a zigzag way by only moving alternate fence posts, but a long, narrow fence with a straight front is preferable. Fences should normally be moved when the cattle are out of the field.

Inputs of fertilizer nitrogen should be similar to those for other systems of grazing, but rest periods can be controlled with greater precision. Strip grazing will produce 12,000 to 15,000 kg milk per ha, which is superior to the output from continuous grazing, but in experimental work strip grazing has not produced significantly more milk than good rotational grazing at the same stocking rate.

Strip grazing is a particularly valuable technique when pasture herbage is in short supply, and the available grass must be rationed carefully. It is equally valuable for use with crops such as kale (Chapter Seven), turnips and rye.

Leader-and-follower System

Because of the selective grazing behaviour of the cow, it is possible to increase milk production slightly by dividing the herd into either two or three equal groups of different milk yields, and graze the paddocks by each group in turn. On day 1, the high-yielding cows consume the highly digestible tips of the grass, whereas on days 2 and 3 the low-yielding animals consume the poorer grass which is left. This leader-and-follower system certainly works, but the trouble of dividing the herd and moving two or three separate groups of cows to the paddocks twice per day has to be balanced against the small increase in daily milk yield. Complications in grazing systems are best avoided, and in most commercial herds the leader-and-follower system is hard to justify. At times of herbage shortage, it is a technique which can help to maintain the milk yields of the high-producing cows but in general it is not recommended. A similar system for calves and young stock is much more worthwhile (Chapter Sixteen).

Types of Fencing

Control of the animals and the correct management of the grass is ensured by sound effective fences. For permanent boundary fencing either a stockproof hedge or a conventional post-and-wire (plain or barbed) fence is required. Fencing posts, 75×75 mm, are adequate at intervals of 2 m with either two or three wires. Semi-permanent fences, which may be in position for 3 to 5 years, can be made from lighter posts, 50×50 mm at intervals of 8 to 10 m with a single electric wire at a height of 750 mm above ground level. Movable electric fences normally comprise metal posts with a foot, and an insulator to carry the electric wire. The electric wire should be about 750 mm above ground level, with the metal posts at intervals of 10 to 20 m. The height of the wire should normally be two-thirds of the distance from the ground to the top of the shoulder of the grazing animal. To economize on the number of posts and to make a stockproof barrier it is essential to strain the wire tightly.

Battery-operated electric-fence units are highly effective if they are kept in good repair with new batteries. Mains-operated units have advantages for both temporary and permanent fences if the farm is compact in shape, and the wire does not have to cross a public road. The importance of an efficient electric fence on the intensive dairy farm cannot be overstressed, and the cattle should be trained to respect it at all times. The electric fence brings a high degree of flexibility into many systems of grazing management, and it is a vital piece of farm equipment. The efficiency of an electric fence can be tested quickly by touching it with a blade of grass, which should be done daily as a routine. Plain wire and twisted wire with a diameter of 1·6 to 2·0 mm, and wire spun with polythene are all efficient, but barbed wire should be avoided.

Great care must be taken when handling electric fence equipment near an overhead electricity supply of high voltage. If the fence wires and posts are too near the cables, accidents can occur, and fatalities have been recorded.

Water Supply

A supply of clean drinking water should be available to dairy cows at pasture at all times. Water is a cheaper commodity than milk, and it is false economy to restrict the water intake of cows in full milk. Reliance on natural water such as streams may be risky, and a plentiful supply of piped water is preferable. Cows eating herbage with a low dry-matter content may consume 13·0 kg dry matter per day plus 50 to 65 kg water in the grass, but to satisfy the total water needs of the animal some extra water is required.

A typical Friesian cow yielding 20 kg milk per day on pasture alone and receiving no concentrates will drink about 40 kg water per day. Water intake increases as the dry-matter content of the herbage increases and as the air temperature rises, but it decreases with increasing rainfall. The peak demand for water could thus arise with high-yielding cows, eating grass of high dry-matter content on a hot and dry day, and the supply of water should be able to meet this peak demand. If this peak daily demand is an average of 50 kg water per cow, then a herd of 80 cows will require a total of 4,000 kg per day. The peak demand for water is in the 2- to 3-hour period after milking time, and the water supply must be adequate at this time. It is advisable to have a high-pressure water supply, pipes and valve inlets with a minimum bore of 19 mm, and sufficient, well-sited troughs of ample capacity. A water trough in the collecting yard is particularly useful, but all fields and paddocks should have a trough with a capacity geared to the size and demands of the herd. If water pressure is low, large troughs holding up to 1,400 kg are needed to act as reservoirs. Sufficient trough space is required for about 10 per cent of the herd to drink at one time. Troughs should be sited away from gates and the corners of fields, and preferably at a point which the cows will all pass.

In large fields where continuous grazing is practised, more than one water trough should be provided. For example, 80 cows on a 16-ha field should have two troughs, so that no animal has to walk too far for a drink. Water bowls connected to either a pipe or a mobile water tanker have a temporary but limited value, troughs of large capacity being preferable. Providing an adequate water supply for a large dairy herd is not a cheap item, but it is vitally necessary for efficient milk production at grass.

Extending the Grazing Season

Depending on the district and the weather, grass is normally available for grazing for about 5 to 7 months per year. If grass can be produced earlier in spring, and later in autumn, the costs of milk production should be reduced, because herbage grazed *in situ* is cheaper and of higher feeding value than conserved grass.

Early grass, often termed 'early bite', can be grown by top-dressing a sward of early perennial ryegrass or Italian ryegrass with about 80 kg nitrogen per ha in early spring. This herbage should be available before the remainder of the grass on the farm, and will serve as an introduction to the main grazing season. If indoor food is scarce, an early bite is valuable but it also presents problems. The response of an early grass crop, when measured in terms of dry matter per kg nitrogen applied, is low, and an early bite can be twice as expensive in fertilizer cost as grass produced later in the season. The early grass also has a low dry-matter content, and thus the intake of nutrients may not be as high as expected. In addition, the grass may be low in magnesium and contribute to the problem of hypomagnesaemia ('grass staggers'—Chapter Seventeen). This risk can be aggravated if the weather is cold and wet, and the cows on the early bite are under stress. On heavy soils in wet weather, an early bite can be severely 'poached' and the sward damaged for the remainder of the season.

Finally, there should be no gap between the end of the early bite and the start of the main grazing period. All these difficulties must be weighed against the advantages of the earlier grazing, which must be rationed carefully each day with an electric fence. A small area of early ryegrass may fit the grazing rotation on a few dairy farms, but on most it is preferable to provide slightly more silage for eating in spring, and to start grazing when all the grass is growing.

Autumn Grazing

Extra herbage can be produced in the autumn by the use of fertilizer nitrogen, but, as in the spring, the response is about half that found in the main growing season. Thus, at the present ratio of milk prices and fertilizer cost, the optimum use of fertilizer nitrogen is about 30 kg per ha in late season. This fertilizer, plus the residual nitrogen in the soil, can provide an additional crop of leafy herbage for grazing in October and November. If the weather is relatively dry, this herbage can supply a useful source of dry matter, although its intake by the cow will not be as high as in the spring. Strip grazing is usually the best method of utilization to avoid undue waste, and poaching should be avoided as this will encourage the ingress of poor grasses and other weeds into the sward at a later date.

If the weather is dry and warm, most cows can make use of autumn grass, although animals in the early stages of lactation will require adequate concentrate feeding for virtually all their milk production. Once the weather conditions become inclement, it is wiser to graze the autumn grass with only low-yielding and dry cows and young stock. The intake and the quality of grass in autumn should not be overestimated.

If swards in certain sheltered fields are dressed with fertilizer nitrogen in the August to September period, and rested until December, a coarse growth termed 'winter foggage' can be produced. This material has little direct part to play in milk production, but it can be an additional source of fairly cheap dry matter for non-milking stock. Foggage production is not of major importance in present-day milk production, but could be of use in certain situations.

Without doubt, it is possible to extend the grazing season in both spring and autumn, and on some soil types it may be desirable, but generally it is wiser to make the maximum use of the grass in the main growing season, and to rely on conserved grass and other crops when grass is not normally growing. The fertilizer nitrogen will then be used more efficiently at periods of optimum response, and the cows will not be subjected to extra changes in feed at times of the year when weather conditions are liable to cause stress.

Supplementary Feeding

The feeding of concentrates to dairy cows at pasture is widespread in Britain. At first sight, this would appear surprising, since leafy herbage has a high feed value (12·1 MJ per kg dry matter) and intakes of dry matter can be around 3 kg per 100 kg liveweight. Thus milk production from grass alone should, in theory, be no real problem if herbage quality and intake could be maintained. Daily milk yields of 20 to 28 kg per cow can be produced from grass alone, but if herbage quality and intake decline, milk yields will decline also. To make up for these two deficiencies, which tend to increase as the season advances, supplementary concentrates are used.

Experimental evidence has shown clearly that when adequate amounts of quickly-grown leafy herbage are available to the cow, there is a low response in milk production to extra concentrates. For example, the mean response from 26 experiments showed that 31 kg of concentrates were required to produce an extra 10 kg of milk, which is highly uneconomic in the short term. At high rates of stocking and with less herbage available, the response will narrow to 12 to 14 kg concentrates per 10 kg extra milk, but this is still uneconomic. The poor response to the concentrates is due mainly to the substitution of the grass by concentrates, and thus the total intake of nutrients is not increased to the extent which might be expected. Supplementary feeding will invariably increase milk yields, but this response requires to be accurately costed. It is always preferable to produce grass of high digestibility and ensure that it is acceptable to the cows, rather than produce inadequate amounts of low-digestibility grass and rely on expensive concentrates.

Cows which calve in autumn should rarely require any supplements at grass until around the time of calving. Cows calving from January to March will produce 20 to 25 kg per day on spring grass, declining to about 10 to 12 kg per day in September. This is a normal rate of decline in yield, about $2\frac{1}{2}$ per cent per week, and any swifter fall can be arrested by either improving grassland management or, as a last resort, feeding a 'buffer' food such as concentrates. A poor response to concentrates must be expected, and the feeding of 4 kg of barley, dried beet pulp, or a cube containing 12 to 14 per cent crude protein will not automatically give an extra 10 kg milk. Any concentrate feeding should be varied according to the grazing conditions and the month, and a simple guide to supplementary feeding at pasture is shown in Table 4.1.

Table 4.1. Rates of concentrate feeding at pasture (kg per 10 kg milk)

Grazing conditions	April, May	June, July, August	September, October
Good	0·5	1·5	2·5
Medium	1·5	2·5	3·5
Poor	2·5	3·5	4·5

(From J. D. Leaver, West of Scotland Agricultural College, 1983)

If the milk yield of cows can be maintained by concentrate feeding, there is the possibility of a long-term benefit, but this response should be costed carefully.

Important exceptions to this overall policy of restricted concentrate feeding must be noted. In early spring when the cows are grazing the first-grown herbage, it is wise to feed certain selected cows 1 to 2 kg per day of a concentrate supplying 60 g of calcined magnesite in order to prevent hypomagnesaemia (Chapter Seventeen). The animals at greatest risk are old cows which are newly calved, and those producing high yields of milk. If at any other time in the grazing season there is a severe shortage of grass, then concentrates will have to form part of the supplementary ration. In general the minimum amount should be given, and the response costed, with forage and forage crops freely available.

Hay and straw offered at the rate of 1 to 2 kg per cow daily on lush spring grass will tend to correct a decrease in the butterfat content of the milk, and at other times these foods may help to prevent the onset of bloat.

Stocking Rate

The importance of the efficient utilization of pasture has been mentioned regularly in this chapter, and a dominant factor influencing this is the stocking rate. In studies of grazing systems it is abundantly clear that an increase in stocking rate will often depress individual animal output, but markedly increase production per ha (Fig. 4.2). If stocking rate is increased too far, both production per cow and per ha will fall. A common fault on many farms is a low rate of stocking in spring followed by overstocking later in the season. About half the annual production of grass occurs by mid June, and therefore stocking should be high during that period. If stocking rate is low, there will be wastage of grass with large rejected areas around the dung pats, and the need to 'top' the pastures. In an intensive grazing system stocking rates should be about 6 cows per ha up to silage time. After this time the rate of stocking can be reduced to 3 to 4 cows per ha.

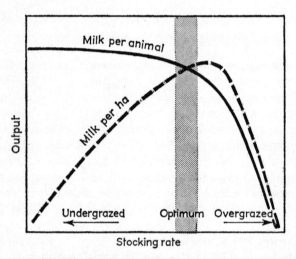

Fig. 4.2.   Relationship of milk per animal and per ha to stocking rate (Adapted from G. O. Mott, *Proceedings*, 8th International Grassland Congress, Reading, 1961, p. 606)

Mechanical Treatment

Without doubt, the most important farm machine on grassland is the fertilizer distributor. Harrowing of grassland in early spring will tear out some dead herbage and scatter molehills, but for the rest of the grazing season harrowing has no value. Rolling fields prior to cutting can press in stones and prevent damage to expensive forage harvesters, but has little effect on the grazed sward. 'Topping' pastures with a mower or similar machine achieves little at most times beyond improving the appearance of the sward by removing tall weeds and clumps of herbage, and tends to retard regrowth. The important time to top pastures is in late autumn, in order to remove clumps of herbage which may cause winter kill.

Zero Grazing

This is a system in which the herbage is cut in the field and carted to the housed animals, and in theory is the most efficient method of utilizing the herbage. Where grassland is not accessible to the stock, e.g. because of a busy road, or because of the large size of the herd, there may be some limited use of this technique, but the system is rarely practised in Britain. A few exponents of zero grazing have been highly successful, but most experimental studies have shown slightly lower outputs per cow and per ha from zero grazing than from normal rotational grazing. The main reason for this difference is that zero-grazed cattle have no chance to select the better parts of the herbage and intakes of dry matter and digestible nutrients are lower than at pasture, where selection is possible. Practical difficulties which also detract from the system include labour demands, mechanical problems with the equipment, contamination of the cut grass in wet weather, and disposal of the slurry which is produced every day of the year. Advantages with zero grazing include savings on fences and water supplies, and the fuller use of expensive harvesting equipment.

A future development could be towards a system of storage feeding, in which silos would be filled with herbage at the grazing stage, and the silage was offered daily to the housed cows. Forage quality would be more uniform than with zero grazing and the field work could be concentrated into the periods when the grass was of optimum quality. However, stocking rate and the yield of herbage would both have to be high for the system to be effective, and with even the most efficient method of conservation there would be some loss.

Target Milk Outputs

As indicated in Chapter Three (Table 3.4), an approximate, but nonetheless useful guide to the potential yield of milk per ha is to assume that every kg of grass dry matter will produce 1 kg milk. The yield of milk per ha from numerous grazing experiments closely parallels the yield of dry matter per ha if the rate of stocking is also increased. Thus, as outlined in Table 3.4, grass with no clover and no fertilizer nitrogen may only produce 3,000 kg milk per ha, whereas the inclusion of a vigorous clover in the sward can lift this value to 6,000 kg per ha. If 125 kg fertilizer nitrogen is applied in addition, the output can reach 8,000 kg per ha, and if 360 kg nitrogen is applied, the yield can be 11,000 kg milk per ha. These target outputs will vary according to the season, the date the cows calve, and the system of grazing, but they are useful values at which to aim. As stated in Chapter Three, an application rate of 375 kg nitrogen per ha can be economic if the herbage is utilized efficiently for milk production, and the present ratio between milk price and fertilizer cost is maintained. The optimum level of nitrogen application for a particular farm will be determined also by other factors and constraints.

Further Reading

Armstrong, R. H., Banks, C. H. and Gill, M. P., *A guide to electric fencing relating to principles, installation and safety*, 1981, Hill Farming Research Organization, Penicuik, Midlothian

Electric fencing, Bulletin No. 147, 1976, Ministry of Agriculture, Fisheries and Food, London

Holmes, W., 'Grazing management', Chapter 4, in *Grass, its production and utilization*, 1980 (ed. W. Holmes), The British Grassland Society, Blackwell Scientific Publications, Oxford

Leaver, J. D., 'Utilization of grassland by dairy cows', *Principles of cattle production*, 1976, Butterworths, London, p. 307

Le Du, Y. and Hutchinson, M., 'Grazing', Chapter 3, *Milk from grass*, 1982 (eds. Thomas, C. and Young, J. W. O.), I.C.I. Agricultural Division, Billingham and Grassland Research Institute, Maidenhead

CHAPTER FIVE

Hay and Dried Grass

Hay Making—Barn Hay Drying—Hay Quality and Feeding—Hay 'Condition'—Types of Hay—Artificially Dried Grass—Dried-grass Feeding—Dried Grass and Silage

Conserved grass forms an essential part of the winter ration of dairy cattle and young stock. In addition, conservation is an integral and important part of the overall grassland management plan. Grazing and conservation are separate techniques of grassland management,

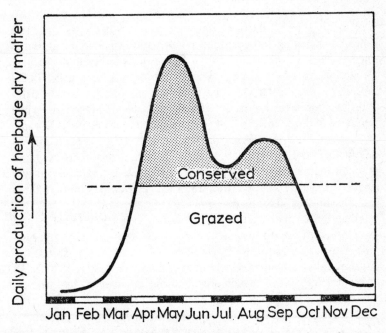

Fig. 5.1. Daily production of well-fertilized grass throughout the year, indicating the relationship between grazing and conservation
(Based on *Forage conservation and feeding*)

but their careful integration benefits both the summer and winter feeding of the stock.

The growth of grass is not uniform throughout the season (Fig. 5.1), and even the most careful selection of early and late grasses and the skilled use of fertilizer nitrogen will not even out the pattern of growth. A peak of grass production occurs in May and June, followed by a trough in mid-season and a smaller peak of growth in the autumn. As the demand for grazing is fairly uniform throughout the period from April to October, the peaks of growth are removed and conserved for use in the winter feeding period.

Two main processes are used to conserve grass. The first involves drying the crop either rapidly, using artificial heat, to make dried grass; or slowly, using natural methods, to make hay. The second process is silage making (Chapter Six) which depends on the controlled fermentation of the wet crop.

Hay Making

Hay making, particularly on farms with small herds, is the most widespread method of grassland conservation in Britain, although on average it produces a food of lower digestibility and nutritive value than that of grass silage. The process of making is misleadingly simple; the crop is cut, dried naturally by sun and wind in the field, and then stored for use in the winter. Most hay fields are cut when the heads of the grasses are fully emerged, and thus the initial D-value of the crop is low. At this mature stage of growth the protein content is low also, and thus the crop is low in both digestible energy and crude protein. Clearly, the quality of hay will be increased by cutting earlier, but yields will be lower, and the leafier crop is more difficult to make into hay in the field.

Thus, traditionally, grass for hay is cut with a low D-value, and the crop continues to lose value throughout every stage in the hay-making process. Respiration in the cut crop lasts from the time of cutting to the moment when the crop is finally dry and in this period the soluble sugars are broken down and feeding value is lost. In hot, dry weather the plants can dry rapidly and losses from respiration may be low, but if the hay dries slowly, losses of up to 15 per cent of the original weight of digestible energy can occur. In addition to this loss, a further 5 to 10 per cent loss can occur because of mechanical damage by the field machinery to the brittle plant, plus 5 to 10 per cent losses in storage due to fermentation. Thus, in total, 15 to 35 per cent of the value of the original crop can be wasted. Methods of speeding the drying process, such as mechanical shaking or tedding of the hay immediately after cutting, and crushing and crimping the cut crop with rollers, are all valuable, but they add to the expense of the hay-making process. These techniques are ideal in favourable drying conditions, but they can increase nutrient losses in wet and difficult conditions.

Baling hay from the swath in the field is an ideal system if the hay is at the correct stage of dryness, 75 to 80 per cent dry matter, and the weather is perfect. In practice, baling is usually done too early, when the crop is not dry. As a result the bale is too dense to allow the moisture to escape, and the hay becomes mouldy, causing problems of both human and animal health. Hay in a store should have a dry-matter content greater than 83 per cent to prevent spoilage and losses.

Hay bales from normal balers can vary in weight from 16 to 32 kg each, depending on the machine and the moisture content of the hay. An average weight of 23 kg is thus vastly different from the 300 to 900 kg bales which can be produced from the big balers. The use of big balers in Britain is increasing, but there is neither evidence nor logical reason why the quality of hay should be improved by their use.

Barn Hay Drying

Methods of increasing the feeding value of hay involve cutting the crop earlier, reducing the period of time from cutting to final storage, and storing the crop at a low moisture content under good dry conditions. These three techniques for improving hay quality can be found in most systems of barn hay drying, but only 4 per cent of all hay is dried in this way. The main advantage of barn hay drying compared with field curing is that the crop is at risk in the field for a shorter period, with the curing conducted under controlled conditions. There is also the advantage of extending the hay-making period at both ends of the season, and taking two or even three cuts during the growing period.

For barn drying, the crop is cut in the normal way but instead of the hay remaining in the field until the dry-matter content has reached 75 to 80 per cent, the hay can be baled at low dry-matter contents of 60 to 70 per cent, and the remaining moisture removed when the crop is in a store and usually under cover. The hay may be only 2 days in the field, and then dried under safe, controlled conditions and put into final store. The hay can either be dried in batches by slightly warmed air and then moved to a final storage place, or dried in the final store by a system of forced ventilation, using hot or cold air.

Specially designed mobile blowers for barn-dried hay have powerful fans which use air which flows over the engine and is thus slightly warmed. If the bales are not too dense and stacked properly, these machines can dry hay efficiently, and the bales will keep in excellent condition over the winter period.

The moving and stacking of bales before drying is not an easy task, and needs to be fully mechanized with bale sledges, elevators and other handling devices. This mechanization, and the cost of the heating and blowing unit, increase the cost of the barn-dried hay but ultimately make a much better product. Mobile machines are available which will dry chopped forage in a heated rotating drum. This system approaches that of dried-grass production in expense and complexity, but does however produce a forage of high feeding value.

In general, if the quality of hay is to be improved, the system of hay making must change from one cut to two or even three cuts per year, with added fuel costs and more complex field and drying machinery. There are considerable managerial, mechanical and drying problems when producing large amounts of barn-dried hay for a large herd, but the technique has a useful place for smaller herds.

Hay Quality and Feeding

Extensive analytical results over many years show that hays have on average a lower D-value and crude protein content than silages, and that barn-dried hays are superior to field-cured hays. These differences between crops occur every year and in all parts of the country. The range of values in both hays and silages is extremely wide, and clearly shows that a high proportion of hays has low D-values. Thus it is typically found that hay feeding makes no contribution to the productive requirements of a dairy herd, and may not even supply nutrients for maintenance. Unless hay has a D-value of at least 60, it must be regarded as of sub-maintenance quality, and unfortunately 80 per cent of hays sampled in Britain are below this value. Hay is rarely offered *ad libitum* to dairy cows, but it is a well-established fact that as the digestibility of hay increases, the voluntary intake also increases. Thus hays with a low digestibility result in low intakes, and the ration must be supplemented generously with concentrates to supply the total nutritive requirements of the cow (Table 5.1).

Table 5.1. Rations for Friesian cows offered hays of various D-values

Hay				Milk yield (kg/cow per day)		
Quality	D-value	ME (MJ/ kg DM)	Food (kg/day)	25	15	5
Very high	67	10·1	Hay	10	17	13
			Concentrates . . .	10	—	—
High	61	9·0	Hay	7	12	14
			Concentrates . . .	14	6	—
Moderate	57	8·4	Hay	6	9	13
			Concentrates . . .	15	8	1
Low	51	7·5	Hay	5	8	13
			Concentrates . . .	16	9	2

With extremely poor hays, of low D-value, it is wise to restrict the intake of hay and offer even more concentrates.

The rations containing hays of different D-values and concentrates shown in Table 5.1 are only a guide if maximum use is to be made of the hay, and assuming that the animals can be grouped and fed according to their yield of milk. The concentrate is calculated to have an ME value of 12·3 MJ per kg DM. Hay is normally offered two or three times per day if 6 kg or more is given daily. If hay is the only forage offered during the winter, the total requirement will be 1·5 to 2·0 t per cow. A deficiency in forage quality can rarely be fully rectified by extra concentrate feeding, and in a season of poor hay, milk yields tend to be lower than in a good hay year. The effect of hay quality can often be seen in the short term when one hay is replaced in the ration by another, and milk yields change rapidly in response to the change in hay digestibility.

Hay 'Condition'

Another factor which influences the feeding value of hay and its acceptability is the combination of characteristics which are described as 'condition'. This is not easy to define accurately. Badly-made hays may be full of dust and smell musty, and patches of mould may be clearly visible. All these factors lower the quality of the hay and its value for milk production. Mouldiness certainly reduces palatability. Mouldy hay is a potential danger to humans, as it can cause the disease 'farmer's lung', and also a danger to cattle owing to its implication in mycotic abortion. A well-made hay smells clean and fresh, with only a little dust to be seen. A green, leafy and dust-free sample of hay is preferable to a bleached, stemmy sample full of dust, and this approximate grading of hays is often reflected in the analysis.

In theory, it would seem that if suitable additives such as propionic acid were mixed with hay, mouldiness would be reduced and quality improved, but on a farm scale additives have not been a success because of the difficulties of their even application in the crop. The use of an additive may incorrectly suggest that hay can be baled at a higher moisture content than usual, but this temptation must be resisted, as early baling will increase the risk of mouldiness and hence spoilage. Hay additives are not an alternative to good drying conditions with sun and wind in the field, and at present they cannot be recommended with confidence.

Types of Hay

Hay is widely classified as either meadow or seeds hay, a distinction which is based mainly on its source rather than on its feeding value. Meadow hay is produced from permanent grassland which is often set aside specifically for this purpose. The swards tend to contain fine-leafed grasses with few legumes, and the hay quality is related to the stage of maturity at the time of cutting. Early-cut hay will be leafy and of high D-value whereas later-cut material will be in full flower and have a low D-value.

Seeds hay is generally made from a temporary ley which may be in use for only 1 to 2 years, and which contains vigorous grasses such as Italian and perennial ryegrass, with or without red clover. Timothy leys, which may last for many years, will also produce high yields of seeds hay but with low D-value and low protein content. The D-value of seeds hay depends on the date of cutting, whereas the protein content varies mainly with the proportion of clover to grass. In general, seeds hay is stemmy with little leaf, and hence of low nutritive value.

Well-made hay with a minimum D-value of 60 is particularly useful for young calves, and for sick animals with a malfunction of the rumen. In general, good hay is a safe feed for all ages of cattle, and has the further advantage of being relatively easy to handle and to feed to stock. There is, however, little place for the low-D-value hay which is so widespread on dairy farms in Britain, and every effort should be made to improve quality by earlier cutting and more rapid curing. This will increase production costs without taking all the risks out of the process, and on the larger farm it is preferable to conserve grass in the form of silage (Chapter Six). This change in the type of conservation crop will require changes in fertilizer policy, machinery, buildings and even in cow housing, but on average a product of higher feeding value will be produced and much of the risk will be eliminated from the vital task of grass conservation.

Artificially Dried Grass

In this conservation technique, the herbage is cut and then transported to a drier, where the moisture content of the crop is reduced quickly to about 10 per cent by blowing heated air through the grass. The drying temperature may vary from 100 °C to 1,000 °C, but the digestibility of the crop will not be reduced if the heat is only used for evaporating the water, and not for heating the grass for a long period. The dried product may be either chopped or made into wafers, cobs and pellets, which progressively reduce the size of the particles in the food. Too fine a particle size is undesirable if the dried grass is the sole feed for dairy cows, as low contents of fat may occur in the milk.

Losses in the field due to respiration and physical damage are extremely small, and if the material is stored correctly the total loss of nutrients from the field to the final product will be only about 5 per cent. This is an extremely low value and is one reason why artificial drying appears to be the best possible method of herbage conservation. Other advantages of drying include the fact that leafy herbage at a young stage of growth, and hence a high D-value, can be conserved, and also that a succession of four or five cuts can be taken over the growing season. High yields of high-D-value dried grass can be produced per ha if sufficient fertilizer nitrogen is applied, but at present the technique is not widespread.

The main use for dried grass and lucerne has been in poultry diets, and dried grass provides only 2 to 3 per cent of the total forage intake of cattle in Britain. The main objections to artificial drying are the heavy expense in both capital and running costs, and the specialized skills required to operate a plant efficiently. Much of the dried grass and lucerne in Britain is produced from large units in the Eastern Counties where grass and legumes provide a useful break in the arable rotation without the need to invest money in animals, fences and buildings. On the small grassland farm in the wetter West of the country, grass drying is not an attractive proposition. Grass drying is technically much more efficient than either hay making or silage making, but the two latter systems involve much less capital expenditure, and can conserve a large quantity of grass in a relatively short time.

Dried-grass Feeding

The nutritive value of dried crops made under good conditions is similar to that of the material before drying, and D-values of 60 to 70 for grass and 55 to 65 for lucerne are normal. For many years, dried grass was assessed mainly on its contents of crude protein and carotene, but a system of grading which includes a crude fibre determination and an estimated D-value and ME content is now practised. The star rating system (Table 5.2) allows 16 possible grades, i.e. 4 D-values at each of 4 crude protein contents. There is a tendency for grass with a high D-value to have an associated high content of crude protein, but this does not always occur. In valuing dried grass for milk production purposes, much more attention should be paid to the D-value than to the content of crude protein, which may have been increased abnormally by a fertilizer nitrogen application.

The different grades of dried grass have different functions in milk production. Dried grass with a 5-star rating, i.e. a D-value of 70 and containing a minimum of 18 per cent crude protein, is almost balanced for milk production, and approximately 1·1 to 1·2 kg of this high-quality food will replace 1 kg of a conventional dairy compound. In theory, as the D-value of the dried grass declines, an increased amount should be offered to replace the dairy compound. In practice the ration would contain an excessive weight of low-D-value grass cubes, which the animals would not eat. This problem of attempting to feed excessive amounts of dry matter would certainly arise with small cows, and with newly calved animals with a high yield (Fig. 2.3).

Table 5.2. Grading system for dried green crops

Star rating	Maximum crude fibre content (%)*	Estimated D-value	Estimated ME (MJ per kg DM)
5-star	17	70	11·0
4-star	21	65	10·0
3-star	25	60	9·0
2-star	29	55	8·5

*In crop with 90 per cent dry-matter content
Note: at each D-value the crop may contain 14, 16, 18 or 20 per cent crude protein
(From A farmers' guide to feeding dried green crops)

Dried grass need not always be considered as an alternative food to concentrates, and in many situations it should be compared on a cost basis (Chapter Two) with hay and silage for part of the maintenance ration of the cow. When assessed in this way, dried grass could often supply nutrients more cheaply than hay, and also have higher intake characteristics. If dried grass is to provide a major part of the winter ration of a dairy herd, it is important to add minerals to the ration to balance any deficiency in the grass. Some straw and low-D-value hay may also be required to maintain the fat content of the milk.

Dried Grass and Silage

High-quality dried grass with a 5-star rating is a useful supplement for high-D-value grass silage. Unlike a supplement of barley, high-quality dried grass given at 3 to 4 kg per 10 kg milk only causes a small reduction in silage intake, and high intakes of dry matter can be achieved in mid-lactation with a production of 20 kg milk per day from two grassland conservation crops. The crops are complementary and indicate the high potential of grassland to produce milk if the two conservation crops are both of high D-value, i.e. 68 to 70.

Dried grass is a source of protein for the dairy cow, but the real economic value of this food will be judged on its ability to produce energy at a competitive price with other foods, particularly barley. In this comparison it is important to remember that barley has a fairly constant composition whereas dried grass has a variable one, and can have a wide range of D-values (Table 5.2). A successful future for grass drying lies in a marked reduction of fuel use, which could be achieved by either pressing the juice out of the crop or wilting the crop in the field by chemical desiccation. If this could be done economically, and the quality of dried crops maintained at a high level, e.g. 68 to 70 D-value, there could be an assured and increasing place for dried grass in dairy cow rations.

Further Reading

A farmers' guide to feeding dried green crops, 1977, British Association of Greencrop Driers, 16 Lonsdale Gardens, Tunbridge Wells, Kent

Connell, J. and Foot, A. S., 'Production and feeding of dried grass wafers to dairy cows', *Biennial Review*, 1972, p. 52, National Institute for Research in Dairying, Shinfield, Reading

Grasses and legumes for conservation, Technical Leaflet No. 2, 1982–83, 1983, National Institute of Agricultural Botany, Huntingdon Road, Cambridge

Hay drying, Farm Electric Handbook, 1979, National Agricultural Centre, Stoneleigh, Warwickshire

Making and feeding better hay, Publication No. 21, 1977, Scottish Agricultural Colleges, Auchincruive, Ayr

Raymond, W. F., Shepperson, G. and Waltham, R., *Forage conservation and feeding*, 1978, 3rd ed. (revised), Farming Press Ltd, Ipswich, Suffolk

CHAPTER SIX

Silage

Silage Fermentation—Wilting—Chopping—Additives—Filling and Sealing—Silage Quality—Silage Feeding—Tower Silage—Big-Bale Silage—Silage Effluent

---

The amount of silage made annually in Britain is increasing rapidly, and on farms with herds of more than 80 cows it is now the main conservation crop. The total weight of herbage dry matter conserved as silage is not as large as that conserved as hay but, because the ME and DCP values of silage are higher than that of hay, the weight of nutrients derived from the two sources is similar. About 70 per cent of all silage made in Britain is used on dairy farms.

The great advantages of silage making over hay making as a conservation technique are that the process is less dependent on the weather, and that the crop can be gathered rapidly so that regrowth is not delayed. Silage making can be more easily integrated than hay making into a system of intensive grassland management, any surplus herbage being conserved rapidly with advantage to the grazing cycle. Advances in mechanization, sealing techniques, and the use of additives have all contributed to the rapid increase in the amount of silage conserved.

Silage Fermentation

The successful conservation of a wet green crop as silage depends on the accurate control of the fermentation process so that the correct acidity develops. Almost every technique in making good silage is aimed either directly or indirectly at the target of acidity control under anaerobic conditions. Briefly, the silage process depends on the fermentation of the sugars in the grass by lactobacilli which produce mainly lactic acid. The rapid initial development of acidity stops undesirable micro-organisms growing so that the crop is virtually pickled in its own juice at a pH of 3·8 to 4·3. Other changes, especially in the nitrogen fractions of the grass, also take place. A high proportion of proteins is broken down to free amino acids, but if the fermentation allows clostridia bacteria to grow, noxious nitrogenous compounds such as ammonia and amines can also develop. In addition, the lactic acid itself may be degraded to acetic and butyric acids, which are not desirable. Satisfactory control of fermentation can be maintained by selecting the correct grass, controlling the moisture content of the crop, excluding excess air, and, where necessary, adding materials which aid the process of fermentation.

Because the level of sugars in the crop affects the initial silage fermentation, the choice of grass and its treatment will affect silage fermentation and hence quality. Ryegrasses have on average twice the sugar content of cocksfoot, timothy and legumes, and the use of a ryegrass for silage making will improve the chance of a good fermentation. Tetraploid ryegrasses contain more soluble sugars than diploid ryegrasses, and will thus improve the fermentation potential of the silage if other factors are equal. The sugar content of grass is lowered by heavy dressings of fertilizer nitrogen, and by dull weather, and is lower in early spring and late autumn than in May and June. A level of 2·5 per cent sugars in the crop as ensiled is generally suggested as a target in order to make good silage. Measurements of sugar can rarely be made in commercial practice and it is wise therefore to use a ryegrass and to cut the crop when dry, if possible in sunny weather.

Wilting

A further way of increasing the sugar content of the crop is to wilt it. Many grass crops contain 17 to 18 per cent dry matter when cut, but after wilting for 1 day with reasonable weather conditions the value can be increased to 24 or 25 per cent. A short period of rain during wilting will not have a major effect on the process, but prolonged, heavy rain, plus dull, sunless weather, will slow down the rate of wilting dramatically. It is advantageous to cut the crop when it is dry, i.e. after the dew and dampness have dispersed. Machines which cut the crop short, such as flail mowers, speed up wilting, but if these machines are used wrongly there can be excessive mechanical losses of grass in the field. Conditioners which crimp, roll, and slightly lacerate the crop can speed up the rate of wilting if the weather conditions are good, but are not effective in poorer rainy weather.

Wilting involves an extra operation, but need not increase the total working time as a wilted crop can be one-quarter lighter than the unwilted crop. Field losses are increased as a result of wilting because the crop is respiring in the swath but there is a reduced production of effluent.

For clamp silage, a dry-matter content not exceeding about 25 per cent is suggested, and, if an additive has not been used, there is an increased chance of a better fermentation than with silage of a lower dry matter content. However, if an effective additive is used at the correct rate there is no advantage to the fermentation as a result of wilting. There is now increasing evidence to show that wilted silage is 7 to 10 per cent less efficient than direct-cut silage as a feed for milk production but it is vital that an additive should be applied to the direct-cut material. Wilting should not normally exceed 24 hours, and this period of time can be reduced considerably if an additive is used. Provision must of course be made for the collection of the silage effluent, which is discussed later in this Chapter.

Chopping

If the crop is chopped and lacerated immediately after it is picked up from the swath, some sap containing the sugars will be released, and thus become more readily available to the bacteria. This favours an improved silage fermentation. The length of silage chop depends mainly on the type of machinery and its adjustment. A flail-type forage harvester will mainly lacerate the herbage, unless the grass has previously been cut and chopped with a flail mower. A double-chop forage harvester will cut the grass into variable lengths from 75 to 150 mm, whereas a metered-chop harvester will give a more precise and shorter chop length from 5 to 75 mm. Flail harvesters can be used either for direct cutting or for picking up a cut swath, and are the cheapest harvesters to buy and to operate. Double-chop machines require more tractor power; they can also be used either for direct cutting or for lifting a cut swath.

The metered-chop machine, also called precision- and full-chop, is the most expensive type to purchase and to operate. This machine cannot cut direct, but has a pick-up reel to lift the grass swath. The short-chopped material gives heavy loads in the trailers and excellent consolidation in the silo. This machine does, however, require the highest tractor power requirement of the various types of forage harvester, and is damaged by stones and metal objects which it may pick up.

As the crop is chopped shorter, the silage machinery becomes larger, more complex and more expensive, and these factors must be balanced against the prospect of obtaining a silage which has an improved fermentation and higher intake characteristics with dairy cows. A relatively new type of machine for silage making in Britain is the forage wagon, which picks up the crop without much chopping and has a low requirement for tractor power and labour. This machine has a restricted use, and, because the grass is not chopped short, the crop must be well consolidated in the silo.

A short chop length of about 25 mm is vital for tower silos and mechanized systems of feeding, but a length of about 50 to 75 mm is acceptable for clamp silage. However, if the herbage has a dry-matter content of approximately 25 per cent, a chop length of 15 to 20 mm is preferable, as this will release more sap from the crop, improve fermentation, increase compaction and make a more acceptable silage for the cattle. The equipment and staff required for some typical systems of silage making are outlined in Table 6.1.

Table 6.1. Some typical systems of silage making

System	Harvester	Staff	Number of tractors	Trailers	ha per 8 hours
Direct	Flail or	2	2	1	1·7
cutting	double-chop	3	3	2	2·7
Wilted	Forage wagon	2	2	0	3·0
Wilted	Flail or	3	3	2	3·7
	double-chop	4	4	2	4·7
Wilted	Precision-chop	4	4	2	5·0

(Based on *Silage*)

Additives

The use of additives to improve silage fermentation may seem unnecessary if ryegrass is cut, wilted, and chopped carefully. Indeed, an acceptable silage can be made without additives, but extensive evidence shows clearly that the use of certain additives is highly beneficial, especially for crops of high digestibility and at times of the year when it may be difficult to obtain a good silage fermentation. An additive is vital with crops of low dry-matter content such as direct-cut herbage. No additive, however efficient, can rectify a basic fault in silage making such as cutting too late or poor sealing, and it is important to improve every aspect of silage making in addition to using an additive. Additives are normally applied when the crop is being cut and lacerated in the forage harvester, and care must be taken with these dangerous chemicals.

Molasses, which contains about 50 per cent sugar, may be used as an additive because the extra sugar assists in the production of acid in the silage. The rate of application is 10 to 20 l per t (litres per tonne) with the highest level being applied to leafy crops of low dry matter content. Molasses is a safe additive, but, because of the relatively high rates of application, handling problems can occur. Application of the undiluted molasses can be either in the field on the cut swath or at the silo.

Acids, in particular formic acid, applied to the grass at a rate of 2·0 to 2·5 l per t, have proved to be highly effective in making an acceptable silage with excellent fermentation characteristics. Extensive studies have shown that formic acid, if used correctly, will reduce the temperature of the silage, decrease wastage and improve the D-value of the crop when compared with silage with no additive. In addition, the voluntary intake of silage treated with formic acid is higher than that of untreated silage, with a consequent increase in milk production. Butyric acid production is restricted, and hence the silage is pleasant to handle, with no offensive smell.

Formalin, which contains 37 to 40 per cent formaldehyde, has some value as an additive, but it is not an ideal material. The level of application is critical, since a low level can promote a clostridial fermentation, whereas a high level can depress silage intake. A further disadvantage is the secondary oxidation, often incorrectly termed secondary fermentation, which occurs when formalin-treated silages are opened and exposed to the air. Formalin can protect the protein from degradation in the silo and in the rumen, and an application rate of 3 to 5 g formaldehyde per 100 g crude protein is proposed as a safe and effective working range. This represents 2·0 to 3·5 l per t as a proportion of formalin to grass with 17 per cent dry matter and 14 per cent crude protein in the dry matter. A mixture of formalin and an acid, either organic or mineral, is more satisfactory than formalin alone, and the recommended application rates are 2·5 to 5·0 l per t.

Other additives include mixtures of acids and sterilants, powders containing lactic-fermenting bacteria, and mineral acids; but the effectiveness of some of these has yet to be shown.

Filling and Sealing

Whatever the system of silage making employed, the silo should be filled as rapidly as possible. Losses in the field should not be high if the crop is wilted for only 1 day, and the rapid filling will exclude the air and keep the temperature low. Filling should be continuous with as few breaks as possible, and if the crop is chopped and an additive used there will be little need for extra rolling and consolidation. During filling, a temporary sheet should be placed over the silo each night to prevent the heated air rising from the clamp and cold air being drawn in. High temperatures are not desirable, and first-quality silage can be made with maximum temperatures of 20 to 25 °C. If the crop is dry and long, it is imperative to fill the silo rapidly.

The final sealing of the silo is a vitally important task and should be done with care and precision. One or more layers of 500-gauge plastic sheeting with generous overlaps at the edges and joints are well worthwhile, and every effort must be made to exclude air and water. The sheet must be firmly held in close contact with the top of the silage by suitable weights such as railway sleepers at the sides, and rubber tyres and bales of straw over the whole surface. A surface pressure of 40 to 50 kg per m^2 and an airtight sealing of the edges are necessary. If air and water are excluded, losses will be reduced and wastage almost eliminated. The construction of silos is described in Chapter Twelve.

The losses which can occur in silage making vary widely. The main sources of dry-matter loss and the amounts are: field losses 2 to 10 per cent, fermentation losses 2 to 10 per cent, respiration in the silo 2 to 20 per cent, effluent 0 to 10 per cent. Thus, excluding the inedible waste at the top and side of the silo, total dry-matter losses may range from about 10 to 40 per cent, with even higher losses of digestible energy. All these losses increase the ultimate cost of each tonne of edible silage and every effort should be made to reduce losses at all stages of the making process so that a total loss of 15 per cent is not exceeded.

Silage Quality

The D-value of silage can vary from about 50 to over 70, and the effect of this variation on milk yield is shown in Fig. 6.1. D-value is a major factor influencing milk production, and on average a change of one unit in the D-value will change the mean daily milk yield by 0·24 kg. Although this amount may appear to be small, the cumulative effect is large, and stresses the importance of improving silage quality if maximum use is to be made of this crop for milk production. Silage with a D-value of 68 to 70 can provide nutrients for maintenance and the production of 12 to 15 kg milk per day, whereas most average silages barely support the requirements for maintenance.

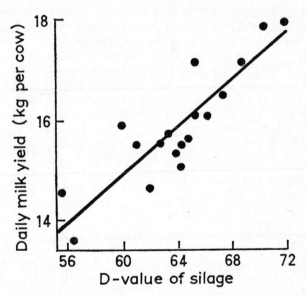

Fig. 6.1. Relationship between the D-value of grass silage, determined *in vitro*, and the average daily milk yield of the cows eating each silage plus concentrates (● represents a specific silage)
(Adapted from original data in *Agricultural Progress*)

As shown in Fig. 3.3, the output of digestible organic matter per ha increases to a peak as the crop matures and cutting is delayed, but the D-value of the silage falls rapidly with a subsequent lowering of its milk-producing potential. Thus if silage cutting is delayed, the yield of dry matter per ha will increase, but extra concentrates will be needed to supplement the poorer-quality silage. A compromise between silage yield per ha and quality suggests a target D-value in silage of about 68 for productive dairy cows. This value can be achieved by cutting at the correct date before ear emergence and by using speedy and efficient silage-making techniques. A high yield per ha can be achieved by using late varieties of grass and applying adequate dressings of fertilizer.

A Friesian cow may eat up to 9 t silage containing 1·8 t dry matter per winter, and this weight of material can be produced from about 0·2 ha intensively managed grass. The first cut of silage provides up to 60 per cent of the total annual output of the sward, but, if the cut is taken early, and the fertilizer applied promptly, the yield of the second cut will be satisfactory. A third cut may be necessary and these later cuts may not have the same high D-value as the first cut. Silage making in most areas should start in mid to late May, depending on the type of grass, the altitude, and the latitude.

Table 6.2. Analyses of high- and low-quality grass silages

	High	Low
D-value	68	52
ME (MJ per kg DM)	10·5	7·6
Dry matter (%)	25	18
Crude protein in DM (%)	17	10
Crude fibre in DM (%)	30	38
Ammonia nitrogen as percentage of total nitrogen	8	25
Ash in DM	8	12
pH	4·0	5·0

Typical analyses of a good high-quality silage and a poor low-quality one are indicated in Table 6.2. The high D-value and ME contents are the prime indicators of quality, but all the other characteristics are of importance. On average a silage with a dry-matter content of 25 per cent is likely to have fermentation characteristics which ensure a higher voluntary intake than a silage of only 18 per cent dry matter. In particular, a low dry-matter silage with no additive will tend to have a high content of ammonia and other products of protein breakdown. Ammonia nitrogen should not exceed 10 per cent of the total nitrogen or silage intake will be restricted.

The pH of the silage is a useful indicator of fermentation quality and a value of about 4·0 is ideal for clamp silage. On rare occasions an extremely low pH may reduce silage intake, but there is no clear-cut relationship between pH and the silage intake of dairy cows. A high level of ash in the dry matter indicates soil contamination, which results in a poorer fermentation and a lowered intake. Many of these characteristics interact, and are further affected by the ferti-lizer and mechanical treatment of the grass and the use of additives. High temperatures in the silo can reduce the D-value of silage by 8 to 10 units. With a cool silage and a good fermentation, the fall in D-value in the silo will be only about 2 units, whereas a poor fer-mentation can reduce the D-value of silage by 4 units. The nutritive value of silage can thus vary widely, and it is extremely useful to have a silage analysis as a guide to feeding the material correctly.

Silage Feeding

In addition to D-value, dry-matter content, and ammonia level, the intake of silage is affected by the amount and type of supplementary concentrate. Barley, low-protein concentrates and sugar-beet pulp reduce silage intake much more than do high protein concentrates. In general, as the protein content of the concentrate is increased, the smaller will be the reduction in silage intake. At an extreme level, the feeding of soya bean meal containing about 45 per cent crude protein will not depress silage intake at all. Even with high-protein silages it is preferable to feed a concentrate containing at least 15 per cent crude protein if maximum use is to be made of silage. Most silages have a high non-protein nitrogen content, and hence a low true protein content, and an adequate level of protein in the concentrates is essential. On average, for each 1 per cent increase in the crude protein content of the supplement, daily milk yield increases by 0·2 kg per cow.

The amount of concentrates to be offered with the silage depends on the choice of the system of concentrate allocation as described in Chapter Two. If concentrates are given according to the daily yield of milk, the amount will be high in early lactation, and decline as the lactation progresses. In early lactation, i.e. approximately the first 10 to 15 weeks after calving, the silage with the highest D-value should be offered, plus concentrates containing 18 per cent crude protein at a rate of about 4·0 kg per 10 kg milk. If a high peak yield is required, reliance should be placed on the concentrates, although large amounts of concentrates may depress the fat content of the milk. As the crude protein content of the concentrate increases, the fat content of the milk will decrease.

In mid-lactation, from weeks 15 to 30, when milk yields will be declining, the intake of silage will increase, and increasing reliance may be placed on it for milk production. Concentrate feeding can be reduced to 2·5 to 3·0 kg per 10 kg milk, depending on the decline in the yield of milk, which should not usually exceed $2\frac{1}{2}$ per cent per week. At this stage of lactation, considerable economies on concentrate feeding can be made if silage is available *ad libitum*.

In the last phase of lactation, i.e. after week 30, maximum reliance should be placed on the silage, and concentrate feeding may be reduced drastically, depending on silage quality. With a silage of high D-value, i.e. about 68, the amount of concentrates should be about 2 kg per 10 kg milk on average over the whole winter-feeding period if the above feeding programme is followed. Silage with a D-value of 68 will sustain average milk yields of 12 to 15 kg per day without any concentrate feeding. Clearly, if silages of low D-value are offered, more concentrates are required as the intake of ME will be lower than with a higher-quality silage.

An example of a typical feeding plan for silage, based on the original idea of Professor W. Holmes, is shown in Fig. 6.2. This illustrates the scheme described previously using a silage of high D-value which is theoretically capable of supplying nutrients for maintenance and the production of 10 to 12 kg milk per day. The concentrates are given at three different levels according to the stage of lactation. Changes in concentrate intake would take place more gently in practice.

Because silages vary so widely in feeding value, it is impractical to give brief tables of typical rations, and in practice a trial-and-error approach based on experience and on the analysis of the silage is necessary. Generally, the value of silage tends to be overestimated, and there is a need for an improvement in silage quality. The protein level in silage can be an incorrect guide to silage quality, and the value of silage should be assessed on its content of ME combined with an assessment of the content of ammonia nitrogen, which should be under 10 per cent of the total nitrogen.

If concentrates are given on a flat-rate system (Chapter Two), the feeding system is much simpler. A fixed daily weight of concentrates, 6 to 8 kg per cow, is given throughout the entire winter-feeding period, and the amount is based on the quality of the silage. For success, the silage must be available *ad libitum* with no undue restriction at either the face of the silo or in the feeding passage. The lactation curve will be flatter than with the previous system of concentrate allocation but total milk yields will be almost identical.

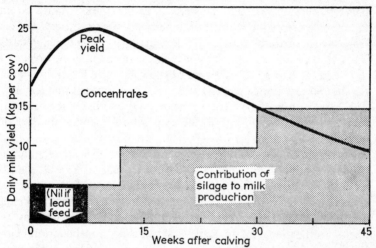

Fig. 6.2. Example of the contribution of grass silage (D-value 68 to 70) to the feeding of a cow calving in the autumn and with a lactation yield of about 5,000 kg
(Based on Professor W. Holmes' original idea, *Agriculture*, 68, 1961, p. 411)

This system of concentrate allocation is simple, accurate, and effective. Some minor modifications are possible such as a mid-season decrease in the daily amount of concentrates, but the essential feature of the system is that all cows are treated in a similar way and that silage is available on a truly *ad libitum* system.

If the quantity of silage available to the cows is limited, other foods such as hay, brewers' grains and roots can be given, but the method of concentrate allocation should still be dependent on the original system selected. Hay is only useful if silage is limited as hay will substitute for the silage dry matter and not increase the total forage and nutrient intake. If roots are offered, silage intake will be decreased slightly but the total intake of dry matter will be increased.

In all systems of feeding, a careful watch must be kept on the body condition of the cow in addition to the yield of milk. Body condition scoring is discussed in Chapter Fourteen; in general cows in mid-lactation should not have a condition score of under 2.

Tower Silage

Tower silos occupy much less area than comparable horizontal silos, but require more specialist machinery such as a precision-chop harvester, dump box and blower. In addition there must be an unloader to empty the tower, and some means of conveying the silage from the tower to the cattle. All this is expensive, and the high costs must be balanced against the convenience of the system. Losses of dry matter in a tower silo are generally low, but field losses are high if the desired dry-matter content of 35 to 45 per cent is to be reached. This high content of dry matter is achieved by wilting in the field and by working the swath, but all these operations increase the loss of nutrients.

To obtain a high dry-matter content, there is also the temptation to delay cutting until the crop is too mature, and as a result the D-value of tower silage is generally lower than that of clamp silage. Grass for tower silos must be chopped to a length of about 25 mm, and filling should be at the rate of at least 3 to 4 m per day to avoid overheating and to achieve satisfactory consolidation. Unloading is done by either a top or bottom unloader; both types operate best with a silage of a dry-matter content of 35 to 45 per cent and 40 to 45 per cent respectively. Concentrates can be added to the tower silage as it is conveyed from the tower to the animals, but the success of this operation depends on the accuracy and efficiency of yet more mechanical equipment.

In the wetter parts of the country, it is not easy to obtain herbage of the correct dry-matter content for making into tower silage, and efforts to achieve a suitable dry-matter content can lead to low D-values. Without doubt, it is easier to make a high D-value silage in a clamp than in a tower if the weather for wilting is difficult. The smooth and efficient operation of tower silos both when filling and emptying depends largely on machinery, which must be well maintained, whereas in clamp silos it is always possible to either self-feed or remove the silage to the cows with simple tractor-powered equipment.

In addition to specialized equipment, tower silage requires some mechanical knowledge and a capacity to operate under a set of fairly exacting conditions. If these are possible and acceptable, tower silage deserves careful consideration on farms in drier areas, where labour is at a premium, the dairy herds are large, and capital cannot yield a better return when used in other ways.

Big-Bale Silage

Making silage in big round bales is a relatively new system in which grass is wilted, baled, sealed in a large plastic bag, and stored carefully until required for feeding. The herbage should be wilted to a dry-matter content of 30 to 40 per cent by tedding the swath, and high-density bales of an even consistency may be achieved by using a baler in low gear at a low speed. The bales should be bagged the same day as they are made in order to establish anaerobic conditions in the bag, and hence a satisfactory fermentation. The plastic bags of at least 400 gauge should be free from holes and carefully pulled onto the bale which is impaled on a tractor fork. A stack of bales may be made on any free-draining, sheltered site which is free from sharp objects and which can be protected from stock. The bales can be stored in the shape of a flat-topped pyramid either three or four bales high, and wide enough to give a square-shaped base. The bags are tied with soft heavy-duty polypropylene twine. Careful tying is vital if air-tight conditions are to be achieved. The bales should be protected from wind damage with either a net or a plastic sheet, and rats and mice must be kept away.

The bales can weigh 350 to 500 kg, and when required for feeding the bag must be opened carefully. The bales are then either placed in a ring feeder or unrolled along a feeding passage. Success with big-bale silage demands as much attention to detail as any other system of silage making. The system is not universally applicable but may act as a logical step in a change from hay to silage making.

Silage Effluent

Silage effluent has extremely harmful effects on rivers and streams, and must under no circumstances be allowed to enter field drains and water courses. A silo containing 300 t of low-dry-matter silage can produce effluent with the same pollution potential per day as a town of 80,000 people, and its damaging effect must not be underestimated. The higher the dry-matter content of the crop at the time of ensiling, the lower will be the volume of effluent, and if a value of 25 per cent dry matter can be achieved there will be almost no effluent.

Effluent should be completely drained away from the silo through glazed pipes and non-porous concrete channels to either a tank or a slurry container. The tank should have a minimum capacity of 3 m³ per 100 t silage and be constructed of non-porous material lined with bitumen. The effluent, mixed with an equal volume of water, should be applied to the grassland at about 18,000 l per ha, when it will supply approximately 20 kg N, 10 kg P_2O_5, and 50 kg K_2O per ha. This rate of dilution is important to avoid scorching the grass.

In conclusion, it is again emphasized that silage is not simply an alternative to hay. A change from hay to silage can alter almost every aspect of the farming system, including the fertilizer policy, the field machinery, the housing of the cows and ultimately the method and place of milking.

Further Reading

Castle, M. E., 'Silage and milk production', *Agricultural Progress*, 50, 1975, p. 53

Castle, M. E., 'Making high-quality silage', Chapter 5, in *Silage for milk production*, 1982 (eds. J. A. F. Rook and P. C. Thomas). Technical Bulletin No. 2, National Institute for Research in Dairying, Reading and Hannah Research Institute, Ayr, Scotland

Hutchinson, A., *Big-bale silage—a practical alternative?* Report No. 30, 1982, Milk Marketing Board Information Unit, Reading

McDonald, P., *The biochemistry of silage*, 1981, John Wiley and Sons, Chichester

Murdoch, J. C., 'The conservation of grass', Chapter 5, in *Grass, its production and utilization* (ed. W. Holmes), British Grassland Society, Blackwell Scientific Publications, Oxford

Silage effluent, Farm Waste Management, Short-Term Leaflet No. 87, 1982, Ministry of Agriculture, Fisheries and Food, London

Thomas, C. and Golightly, A., 'Winter feeding', Chapter 2 in *Milk from grass*, 1982 (eds. C. Thomas and J. W. O. Young), I.C.I. Agricultural Division, Billingham and Grassland Research Institute, Maidenhead

U.K. silage additives, 1983 season, ADAS 1983, Ministry of Agriculture, Fisheries and Food, London

CHAPTER SEVEN

Forage Crops and Concentrates

Succulent Fodders—Turnips and Swedes—Mangolds, Fodder Beet and Sugar Beet—Carrots, Parsnips and Potatoes—Kale—Cabbages and Rape—Root Tops—Cereals—Forage Maize—Red Clover—Grass and Forage Crops—Comparative Yields—Straw—Straw Processing—Concentrated Foods—Energy Straights—Protein Straights—Concentrate Formulation—Fats and Oils—Urea and Biuret—Blocks and Liquid Feeds—Brewers' Grains

Grassland and its conservation crops are the main source of nutrients for the dairy cow (Chapter Three), but a range of forage and root crops is grown also for feeding to ruminants. During the period from 1950 to 1970 there was a marked decline in the area of these crops, a small increase in about 1975, and a steady decline thereafter. In total, fodder crops constitute less than 1·0 per cent of the total area of crops and grass in England and Wales, and this area is not all devoted to the feeding of dairy animals.

Succulent Fodders

The forage and root crops contain 70 to 90 per cent water, and are often termed succulent fodders, in contrast to the dry fodders such as hay and straw, which may contain only 15 to 20 per cent water. Most forage and root crops are of value for only one season, and unlike grass normally do not produce a second crop when they have been utilized by cutting, grazing or harvesting. Such crops can provide valuable food when grass is not available in the winter and spring months (Fig. 7.1). For many years the root crop was the pivot of a successful farm rotation, but now such crops play a much smaller part on the farm, and may even be grown mainly to utilize heavy applications of slurry from the dairy herd (Chapter Thirteen).

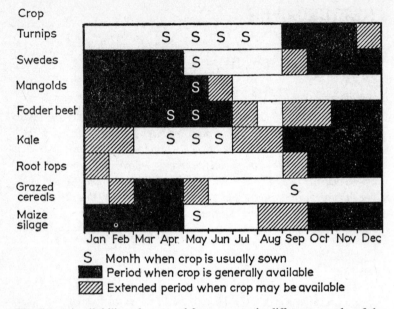

Fig. 7.1. Availability of root and forage crops in different months of the year
(Based on M. E. Castle, in *Animal Health, Production and Pasture*, 1963, Longman, Green and Co Ltd, London)

The succulent fodders may be divided into two main groups according to the part of the plant that is of the greatest importance as a source of food for the stock. Root crops include turnips, swedes, mangolds, beet, carrots, parsnips and potatoes. Forage crops include kale, cabbage, rape and maize, in which it is the leaves and stems which are eaten. There are important nutritional differences between these two groups.

The *root crops* are sources of energy, for although they contain little fat, their contents of sugar and starch are high. In addition, their fibre content is largely cellulose and hemi-cellulose and is therefore well digested by ruminants. Generally, the protein value of roots is low—4 to 13 per cent of the dry matter—and a high proportion of it is in the non-protein nitrogen form. Roots are a poor source of calcium and phosphorus, but are rich in potassium. The dry-matter contents of root crops vary widely, and this measure of their nutritive value is essential both in formulating rations correctly and in valuing the crop. Early-maturing turnips may have a dry-matter content of only 8 per cent, whereas some types of fodder beet may reach a value of 22 per cent. With such a large range in dry-matter contents it is meaningless to discuss yields per ha unless the crop has been sampled carefully and a dry-matter value determined.

The *forage crops* generally have a much higher content of protein than the root crops, and are high in vitamins, particularly carotene, but frequently contain indigestible lignins. The dry matter of forage crops contains less carbohydrates than that of root crops, and usually has a lower D-value. Root crops such as mangolds and swedes may have D-values of around 80, whereas typical forage crops including kale and cabbage may be in the 65 to 70 range.

Turnips and Swedes

Turnips (*Brassica rapa*) have rough hairy leaves, and the flesh of the root is either white or yellow. Early-maturing turnips, because of their low dry-matter content, should be fed in autumn and early winter before the frosty weather arrives. The crop can either be folded with cattle or, more expensively, be pulled with the tops attached and fed on pasture. Hardy turnips, especially the yellows, can be treated and fed as swedes. Care must be taken when feeding turnips to dairy cows just before milking time, as the crop may taint the milk. The taint is due to volatile organic compounds, which may enter the milk either from the bloodstream of the cow or from the air. The crop must not be stored where the animals are milked, and should not be given to the cows later than 2 to 3 hours before milking time.

The sowing of stubble turnips, i.e. Dutch White turnips, into a cereal crop either before or at harvest time can provide grazing in the autumn. A far more predictable method is to sow the Dutch White turnips after a second cut of grass silage. The turnips must be sown in July and August, and both the yield of the crop and its utilization are extremely variable. The crop has the disadvantage of a poor root anchorage which can cause high levels of wastage; an average yield of 5 t dry matter per ha often being halved.

Swedes (*Brassica napus*) have smooth glaucous leaves and a short neck above the root, and are usually yellow-fleshed. They have a dry-matter content of 9 to 12 per cent. The early-maturing swedes with a low dry-matter content should either be used before winter or stored in clamps. The late-maturing varieties with a high dry-matter content can be left in the field and lifted as required for the stock, even in January and February. Swedes are rarely grazed by dairy cows, and if the crop is stored carefully in clamps and sheds, losses can be as low as 5 per cent. In general, swedes are a safe and valuable crop and can yield more dry matter, digestible organic matter and protein per ha than barley (Table 7.1). Swedes can taint milk, but not as badly as turnips. The feeding value of turnip and swede dry matter is similar and can replace cereals and hay in the ration. Swedes should be introduced into a ration gradually over a period of 1 to 2 weeks and may be offered either chopped or whole. There are claims that chopping has little beneficial effect, and systems of feeding whole roots in large wooden hoppers have been developed with success.

Table 7.1. Yields of dry matter and digestible organic matter from some forage and other crops

Crop	Dry matter	Digestible organic matter	Crude protein content of dry matter (%)
		(t per ha)	
Fodder beet	10·0	8·0	7·0
Maize	10·0	6·5	9·0
Kale	9·0	6·5	15·0
Red clover	8·0	4·0	19·0
Swedes	7·0	5·5	11·0
Stubble turnips	5·0	3·5	14·0
Rape	4·0	3·0	20·0
For comparison			
Barley grain	4·5	4·0	11·0
Ryegrass	9·0	6·0	17·0

These hoppers, termed ADAS-Walcot feeders, hold 2 to 3 t of roots which are self-fed by gravity to the cattle. In many ways, a swede crop is a direct rival to the silage crop, and if swedes are fully mechanized, i.e. by precision sowing, inter-row cultivation, herbicide treatment, harvesting and feeding, they could have an increasing part to play in feeding dairy cows.

Swedes and other root crops have the important characteristic in dairy rations of increasing the voluntary intake of dry matter when incorporated into a diet of dry feeds and silage. This is attributed to the rapid breakdown of the roots in the rumen and their low content of indigestible crude fibre. Thus, the addition of 1 kg of root dry matter to a ration will only reduce the forage dry-matter intake by approximately 0·5 kg, the net result being an increase of 0·5 kg dry matter. This method of increasing the supply of energy to the cow can be particularly useful in early lactation, but an adequate supply of protein must be assured. Swedes contain about 11 per cent crude protein in their dry matter, but 40 per cent of this is in the form of non-protein nitrogen and there is a possibility that with some rations there may be a shortage of protein in the small intestine.

Mangolds, Fodder Beet and Sugar Beet

These three crops are all subspecies of the same genus (*Beta vulgaris*) but have been bred for different purposes. Mangolds and fodder beet are for stock feeding, whereas sugar beet is grown for the extraction of sugar for human use, the root seldom being available for direct feeding to stock.

Mangolds have the lowest dry-matter content, with values ranging from 9 to 15 per cent. They are a valuable source of carbohydrates, but contain a high proportion of non-protein nitrogen. Some of this nitrogen is in the form of nitrate, and the freshly-lifted crop can cause scouring in livestock. However, the problem can be avoided by lifting the crop and storing it for use after Christmas. Mangolds can be given to dairy stock in exactly the same way as swedes, and cause no taint in the milk.

Fodder beet is a general term used to describe a wide range of roots of varying characteristics. Briefly, the high-dry-matter fodder beets may have dry-matter contents of 18 to 22 per cent, and often resemble sugar beet in such characteristics as colour and depth into the soil. The medium-dry-matter fodder beets have dry-matter contents of 14 to 18 per cent and are similar to the mangold crop, growing relatively high out of the soil. It is again emphasized that a knowledge of the dry-matter content of the crop is vital in assessing and correctly feeding fodder beet. The feeding value of mangold and fodder-beet dry matter is similar, and one crop can be substituted for the other on a dry-matter basis. Fodder beet is an extremely palatable food for cattle of all ages, but care must be taken to introduce it gradually into the ration to avoid digestive upsets. This problem is probably linked with the high sugar content of the fodder beet, but it may also be associated with an excessive intake of soil on the roots.

All root crops should be reasonably clean when given to livestock.

Sugar beet which is not required by the factory may sometimes be available for cattle feeding. This crop has a dry-matter content of 22 to 25 per cent, with up to 75 per cent of sugars in the dry matter, and it can be given to cattle in much the same way as fodder beet. Care in feeding is, however, vital, and it is essential to introduce the crop slowly over a period of about 10 days.

With all root crops, a mean daily intake of about 3 kg dry matter per cow is a safe and reasonable amount, but this quantity can be varied widely, depending on the availability of the crop and the relative price of alternative feeds.

Carrots, Parsnips and Potatoes

Crops such as carrots, parsnips and potatoes are rarely grown for feeding to livestock, but at times they may be available at a price competitive with other stock foods. Carrots have a dry-matter content of about 13 per cent, and can be safely given to dairy cows, with benefit to the colour of the milk on account of their high carotene content. Parsnips may contain 15 per cent dry matter, and may also be given to cattle so that they replace swedes on an equal dry-matter basis.

Potatoes differ from other root crops in containing about 70 per cent starch in the dry matter, with only a low content of sugars. The average dry-matter content may be about 23 per cent. Raw potatoes must be offered to cattle with care, and be slowly introduced into the ration. They should be clean, and preferably sliced to guard against choking. At times, cows may bloat on potatoes and a daily ration of 2 to 3 kg dry matter per cow is a safe maximum. Diseased and green potatoes and any green sprouts must be removed.

Kale

The kales (*Brassica oleracea*) contain a wide range of plant types, the most commonly grown being Marrow-Stem kale, Thousand-Head kale, Hungry-gap kale, rape kale, and an outstanding hybrid kale called Maris Kestrel. This variety is an ideal kale for most purposes, having a high yield and a high feeding value combined with winter hardiness. Kale can produce a high output of nutrients per ha if well cultivated and adequately fertilized, but its utilization is often poor.

Kale has a dry-matter content of 14 to 16 per cent, and contains about 15 per cent crude protein in the dry matter. In addition, this crop has a D-value of about 70 and contains 1 to 2 per cent calcium in the dry matter. Kale has thus a high potential feeding value for the dairy cow, and can fill a gap in the feeding programme after the grass is finished in the autumn (Fig. 7.1). In spite of all these nutritional advantages, the area of kale has declined rapidly since 1970, and this reduction is related to its utilization and also to the problem of cropping the kale stubble after ploughing. Years ago, it was common to cut kale by hand and cart it to the stock either in a building or at pasture. At the present time, when it is important to make the maximum use of expensive labour, this unpleasant task has been eliminated by strip-grazing the crop (Chapter Eight).

To obtain the maximum output from the land, and to improve the utilization of the kale when it is grazed, the following technique of direct drilling can be used. Briefly, a grass sward which has yielded a crop of silage in May or early June is killed by an application of paraquat and the kale seed is drilled directly into the dead grass sward. If the seed is sown before the end of June, there is normally sufficient moisture in the soil for a good establishment of the kale. Then, with an application of 100 to 120 kg fertilizer nitrogen per ha, a crop of about 8 t dry matter per ha can be produced. Excessive yields are not required, as wastage can be high with tall crops.

A disease termed kale anaemia, which is characterized by a fall in milk yield, dullness and even blood in the urine, can occur under some conditions if kale is fed in excess. To prevent this problem, kale should not exceed one-third of the total dry-matter intake of the cow, and since the toxic substance increases from August to January, kale intake should be reduced in the late winter. The flowering heads of kale should not be grazed.

In practice problems with kale are not widespread, and this important crop can make a valuable contribution to feeding the dairy herd economically at a difficult time of the year.

Cabbages and Rape

Cabbages are a safe and palatable food for cows, the two main types being drumhead and open-leaved. The dry-matter content of cabbage is lower than that of kale and may be only 11 per cent and even lower, but the D-value is about 67. The crop may be cut and carted to the stock, but grazing behind an electric wire can be successful on dry soils. Cabbages can be offered in amounts up to 3 kg dry matter per cow daily but, because of the risk of taint in the milk, feeding should be after milking and preferably in the open air. Brussels sprouts and processing waste can also be safely given to dairy cows. Rape is rarely grown specifically for feeding to dairy cows. It is similar to kale in nutritive value, but can taint the milk.

Root Tops

The green tops of fodder beet, sugar beet, turnips and swedes can all be used for feeding cattle, and have a nutritive value similar to that of kale. Care in feeding is required, however, as beet tops contain oxalic acid, which causes scouring and may even kill the animals; tops must be wilted in the field for 1 to 2 weeks before being offered to the stock. Soil contamination should also be avoided, and the tops kept in clean rows in the field. Swede and turnip tops have a crude protein content of about 20 per cent in the dry matter, and are an excellent source of carotene. Sugar-beet tops consist of the leaves and a portion of the crown and have a D-value of about 70, whereas fodder-beet tops are generally all leaf and thus have a lower D-value. Maximum daily intakes of tops should be about 3 kg dry matter per cow and feeding should be after milking. A careful watch must be kept on the cows' dung, and the amount of tops reduced if scouring occurs.

Cereals

The cereal crops—wheat, oats, barley and rye—are grown principally for the production of grain, but a small area is used for grazing. Occasionally cereals are sown with grass and clover mixtures when reseeding grassland direct. The rate of seeding is approximately 40 kg cereals per ha, but the value of the technique is questionable if the conditions for ley establishment are good enough to allow rapid growth of the grass and clover.

Cereals, either sown alone or mixed with a legume such as peas and vetches, are used at times as a green forage crop. The crop can either be cut for feeding direct to the cattle or be made into silage. The crude protein contents and D-values are low, but the carbohydrate content is high, especially when the grain is at the milky stage. This crop has only a minor place in present-day intensive milk production.

Another use of cereals is to sow a crop of rye, variety Lovaszpatonai, in autumn—not later than mid-September—for grazing in early spring. This early bite can be grazed 1 to 3 weeks before the grass is available, and may stimulate milk yields at a time of year when other food might be scarce. Unfortunately, the crop matures rapidly and can quickly become tall, unpalatable and of low D-value. The cost of producing this special-purpose crop with a short season of vegetative growth has to be charged completely to the dairy herd, and it is often more efficient to produce slightly more silage than to complicate a simple farming system with an extra crop. If rye is grown, it must be carefully integrated with a following crop such as Italian ryegrass, and this can limit its use.

Forage Maize

Maize (*Zea mais*) was first considered as a fodder crop in Britain about two centuries ago, but it is only since 1970 that it has been grown on a considerable scale. In 1977 approximately 33,000 ha were sown, but the area has now declined to approximately 15,000 ha in 1982. The pioneer maize growers were mainly dairy farmers with large farms who made grass silage, and it was thus possible for the labour and machinery demands to be spread over two crops at different times of the year. The interest in maize is largely due to the successful breeding of suitable hybrid varieties, its high D-value, and its relatively low labour requirement if full advantage is taken of mechanization.

Forage maize can only be grown in certain areas in Britain south and east of a rough line from the Wash to the Bristol Channel. Even in this part of England, late spring frosts can check the crop, and elevations of over 130 m are to be avoided unless the selected field faces south and is well protected from the wind. It is estimated that a total of 728 °C-days or more are required in order to grow maize with cobs of a milky consistency 9 years out of 10. This value is calculated from the average accumulated temperatures above 10 °C for the 6-month period May to October. There have been optimistic forecasts that maize could be grown with success in specific areas in the North of England and South-West Scotland, but in general prospects are not hopeful unless new varieties are bred.

Maize is an ideal crop for utilizing the nutrients in a heavy application of slurry, since it seldom lodges. The seed is sown in early May when the danger of frost is virtually over, and the aim is a population of about 100,000 plants per ha. Spraying to eliminate weeds is important, and birds must be kept away from the grain after sowing. Early- and medium-maturity varieties are required if the crop is to be harvested in late September and early October. Early-maturing varieties are essential in the more northern part of the maize-growing area if the crop is to be at the correct stage for silage making before mud and rain make conditions difficult for harvesting.

The entire plant is harvested by precision-chop forage harvesters with special maize attachments which chop the maize cleanly into lengths of 10 to 15 mm. Flail harvesters are not suitable. The dry-matter content of the maize should be about 25 per cent if a clamp silo is used, and about 30 per cent for tower silos. Low dry-matter contents cause problems with effluent, whereas high dry matters may result in moulding owing to lack of consolidation. A long clamp silo with a small area of face is advocated to avoid the entry of air, overheating and secondary oxidation when the silo is opened for feeding. The entry of air into maize silage must be prevented, and short chopping, rapid filling, consolidation and perfect sealing are vitally important.

Maize has a high D-value, and one of its important characteristics is that the D-value does not decline as the crop matures from mid-August to October. The proportion of the total crop contributed by the cob increases in this period and thus maintains the D-value. The cob can constitute 50 to 55 per cent of the total dry matter in a maize crop, and this is an ideal proportion for a silage of high D-value and a good fermentation.

A typical maize silage has a dry-matter content of 25 to 30 per cent, a D-value of 65 to 68, and a pH of about 4·0. The content of crude protein is usually low, values of 9 to 10 per cent in the dry matter being normal. Maize silage is also low in all the major minerals. To counteract these deficiencies it is essential to feed adequate amounts of protein and minerals to the cows. Additives containing non-protein nitrogen and minerals are available for incorporating into the silage when the silo is being filled.

Daily intakes of silage dry matter can increase from 7 to 11 kg per cow as lactation advances. In early lactation the contribution from concentrates should be high, with a restricted intake of maize silage, whereas in later lactation the silage can be given virtually *ad libitum*.

The concentrate used to balance the maize silage should contain a minimum of 16 per cent crude protein, but higher contents are justified in early and mid-lactation, especially if the maximum use is to be made of the silage. To maintain the butterfat content at a satisfactory level, the daily provision of 2 kg hay per cow is recommended.

Red Clover

Red clover (*Trifolium pratense*) of both the broad and late-flowering types has been included in seeds mixtures for many years, but recently there has been a renewed interest in certain of the broad types as an arable crop. This interest has been stimulated by the breeding of both diploid and tetraploid varieties which are superior to the older varieties in yield per ha and in persistency. Varieties such as Astra, Norseman and Redhead will persist for 2 to 3 years and produce high yields without the use of fertilizer nitrogen. The clover can be sown either under a cereal crop or direct, and a fine and firm seed bed with a pH of 6·0 to 6·5 is essential for a successful establishment. The seed can be sown alone or mixed with a medium to late perennial ryegrass (Table 3.2, mixture 4). Italian ryegrass is too aggressive and should not be used.

Red clover has a dry-matter content of 12 to 14 per cent, and is not easily made into hay because of the different drying rates of the leaf and stem. The leaf can dry rapidly and shatter, with a serious loss of valuable nutrients. It is preferable therefore to conserve the crop as silage, which can be done satisfactorily if the clover is wilted to a dry-matter content of about 25 per cent, chopped into lengths of 15 mm and formic acid added at 2.5 l per t. Red clover swards tend to be open, and it is important not to pick up any soil with the crop if a correct fermentation is to be obtained. With wilting and the use of an additive, a pH of about 4.2 should be obtained.

The D-value of red clover silage averages about 56 but, because the intake of a legume is 20 to 40 per cent higher than that of a grass of the same D-value, satisfactory intakes of 10 to 11 kg of silage dry matter per cow are possible. Although the crude protein content in the dry matter of red clover silage may be about 19 per cent, it is still advantageous to feed dairy cows a concentrate containing a minimum of 14 to 16 per cent crude protein. The high intake characteristic of red clover silage is one of its most valuable features, and there can be occasions when a mixture of grass silage and red clover silage could be useful to increase intake.

Red clover contains oestrogens which can affect fertility in some classes of stock, but experimental work indicates that the silage can form the complete forage intake of dairy cows for the entire winter without adversely affecting the breeding pattern. Red clover silage is normally only a part of the total winter ration, and the risk of a problem of infertility from this food is remote. Red clover will yield about 8 t dry matter per ha annually over a 3-year period, but its use will depend ultimately on the possible integration of this arable forage crop into the farming system. In many grassland areas there is no advantage in substituting a forage crop for productive grassland, but elsewhere red clover could fit extremely well into the rotation.

Grass and Forage Crops

On grass farms which have a high rate of stocking, and which are relatively small, there is little place for forage crops. If grass is worn-out and requires renewing, this can be done by a direct reseed. Kale is probably the only forage crop worth considering on this type of farm, as the crop can be sown after a silage cut has been taken. Root crops such as swedes can be purchased if the cost of the nutrients is competitive with that from other crops, and this may be preferable to growing a crop which requires specialized machinery and high peak labour demands.

On larger dairy farms there is usually some cropping to produce barley for feeding, and some of this cereal area can often be better employed in growing a root crop. The yield of dry matter and nutrients per ha is higher from a root crop than from an average cereal crop, and roots included in the winter ration can increase voluntary feed intake. Land for forage crops such as roots and maize is an ideal place for heavy dressings of slurry and dung, and is preferable to grassland for this purpose.

On the large arable farm, forage crops and by-products such as sugar-beet tops and vegetables can form an important part of the dairy cows' intake. Grass may be grown purely as a break crop with the dairy herd as a useful and profitable means of utilizing it. In suitable situations in the South, maize is the ideal forage crop if the farm is highly mechanized and has high-powered tractors. Some rotational constraints may have to be imposed on some brassica catch crops and fodder beet, as these crops will act as hosts for eelworm, which affects the sugar-beet crop.

Comparative Yields

Some estimates of yields of dry matter and digestible organic matter per ha from a selection of forage crops sown in suitable environments are indicated in Table 7.1. This type of yield information is only an approximate guide, and the output of a specific crop has to be considered in relation to the district, soil fertility, season and management. In addition, the ultimate value of the crop depends on the proportion of it which is utilized by the animals. For example, a badly grazed crop of kale may have 40 per cent waste and only 60 per cent utilized. In contrast, a well harvested and stored crop of swedes may be 95 per cent utilized. Nonetheless, it can be seen that maize and fodder beet are capable of high yields of dry matter per ha, and can outyield ryegrass receiving 300 to 350 kg nitrogen per ha. Kale, maize and fodder beet all produce high yields of digestible organic matter per ha. The outputs of crude protein per ha are not given, as the proportion of non-protein nitrogen can vary so widely both within and between crops, and thus give misleading data.

The output of nutrients per ha is only one criterion in selecting a forage crop; other important considerations are the availability of the crop at different times of the year (Fig. 7.1), and the cost per t of producing digestible organic matter. Crops with low labour demands such as grazed kale can produce cheaper nutrients than crops demanding either much hand work or expensive specialized machinery. The choice of forage crop can also be dictated by the special requirements of the cow at different stages of lactation. A herd calving in spring will make excellent use of kale in the autumn when milk yields are fairly low; on the other hand, cows newly calved in autumn will be cleaner and more comfortable at a self-feed silo face than eating kale in a field. Cows with high yields in early lactation require the maximum intake of nutrients, and should be spared the stress and loss of energy caused by walking and grazing in a cold environment. Offering swedes in a covered feeding passage may be preferable to eating stubble turnips in a muddy field.

Finally, in selecting crops for feeding to the dairy herd, it must be remembered that the area of land used for these crops may cause a restriction on either the grazing or the grass conservation areas. In summer this can cause problems on a small farm, and it may be necessary to feed some of the forage crop. For example, maize can be cut and given to the herd, but this of course depletes the winter feed. On a large farm which is not heavily stocked this problem is not as noticeable, but on a small farm it is impossible to gain forage without depleting grass. Well managed grass can be an extremely high-yielding crop, and any alternative forage crop which is grown must have some clear-cut advantage over the grass crop.

Straw

Straw consists of the fibrous stems and leaves of plants after the seeds have been removed. During the ripening of the cereal crop, nutrients are transferred from the stem to the seeds, and hence straw is of low nutritive value. The D-values of straw are in the range of 40 to 45, and the contents of crude protein and phosphorus are low also. Voluntary intakes are generally low, and thus straw makes only a small contribution to the total nutrient intake of dairy cows.

Straw Processing

Chopping and laceration of straw will increase intake slightly but not affect D-value. If straw is treated with an alkali, such as sodium hydroxide, the indigestible lignin is separated from the cellulose, and both the D-value and the voluntary intake of the processed straw are increased.

Years ago, straw was soaked for 24 hours in a 1·5 to 3·0 per cent solution of sodium hydroxide, drained and well washed to make a wet and neutral product, with a pH of about 7. The final product had an improved feeding value, but losses were high.

The present technique for processing straw is either to mill or chop the straw, and immediately spray it with a 16 per cent concentration of sodium hydroxide to form a dry, alkaline product with a pH of about 11. The rate per t straw is 360 l solution, which weighs 425 kg and contains 68 kg sodium hydroxide. As a result of this treatment, D-values can increase from 45 to over 60, with a resultant increase in the voluntary intake of the processed straw. Although sodium hydroxide is an extremely dangerous and corrosive substance, the processed straw can be handled with safety 15 minutes after treatment.

The alkali-treated straw can be either stored in a silo for future use or fed to stock at once. It is of particular value in complete diets for growing and fattening stock but is of more limited use for dairy cows, having little place in rations for high-yielding cows. Because of the high capital costs of the machinery, the process is more suited for large farms and co-operative ventures.

Alkali-treated straw does not pose any problems to animal health, but cattle offered this food will drink 30 per cent more water per kg dry matter than they would if given untreated straw. As a result, urine output is increased and it is preferable therefore to house the animals on slats. Rations containing processed straw should be adequately supplemented with protein and minerals. Cubes containing alkali-treated straw, termed nutritionally improved straw (NIS), are available commercially and have a limited, but useful place in some dairy rations when alternative foods such as hay and silage are scarce. A typical analysis of NIS is: estimated ME 9·0; crude protein 4·5 per cent; and crude fibre 39 per cent of the dry matter.

Concentrated Foods

Although the importance of a high D-value in conservation and forage crops has been stressed in previous chapters, it is necessary to include concentrated foods in the ration of the cow at certain stages of the lactation. These foods are usually low in crude fibre, high in ME, vary in crude protein contents and are described as either concentrated foods or merely concentrates. This term applies to a single food and also to a mixture of foods. A concentrate is more strictly defined as a food supplying a primary nutrient, e.g. ME and containing not more than about 16 per cent crude fibre with a dry-matter content over 80 per cent. There are exceptions to this definition, but in general it covers the cereal grains and by-products, oil cakes and meals, foods of animal origin, and a range of miscellaneous foods. In economic surveys the term concentrates includes all the above foods but often excludes wet brewers' grains, i.e. draff.

The following terms are now agreed within the feeding industry and the advisory services. 'Straights' consist of a single food such as barley, beans or fish meal. 'Compound feed' is a mixture of straights plus minerals and trace elements to provide a balanced food. Compound foods may be either purchased from the feed industry or mixed on the farm, i.e. a home mix. 'Protein concentrates' are a product specifically designed for mixing with cereals before feeding. 'Concentrate' is a generic term to include the three previous groups.

Energy Straights

This group of foods include the main cereals—oats, barley and wheat —and their various by-products; also maize, rice, millet, sorghum, rye, dried sugar-beet pulp, fats and oils. The ME values range from about 10·0 to 13·6 MJ per kg fresh weight, with the exception of the fats and oils which have high values and are mainly used by the feed industry in proprietary concentrates. The crude protein contents of the foods in this group are normally low, and rarely exceed 11 per cent of the fresh weight. Some exceptions with high crude protein contents include maize-gluten feed, malt culms, brewers' grains, dried brewers' yeast, distillers' grains and distillers' solubles.

Protein Straights

These foods include the oilseed cakes and meals which are the residues after the removal of oil; they include soya bean meal, cotton-seed, linseed, coconut, and palm kernel. The oil is removed from the original crop either by pressure, termed an expeller process, or by solvent extraction, when the residue is described as extracted. If the food has a husk it is termed undecorticated, and has a lower nutritive value than the decorticated food which has the husk removed. The protein straights can be conveniently divided into those with either a high or a low protein content. Soya bean meal has a high protein content, whereas coconut cake, palm-kernel cake, beans and peas have a low protein content. Some typical analyses are given in Appendix One. Animal proteins such as fish meal, meat meal, and meat and bone meal are included in the group of protein straights and are also a useful source of minerals for cattle. Fish meal has an exceedingly high content of crude protein which is of low degradability (Table 2.6), and hence can be a valuable food for balancing rations for high-yielding cows.

Concentrate Formulation

A balanced concentrate for a specific feeding purpose can be obtained in one of three different ways. Firstly, a proprietary cube or meal can be purchased from the feed industry; secondly, a protein concentrate, commonly termed a 'balancer', can be purchased and mixed with cereals on the farm; or thirdly, the straights can be home-mixed. The proprietary foods are accurately formulated and mixed at a mill and are delivered to the farm either in paper sacks or in bulk. Labour is saved by not mixing on the farm, as is the capital which would be required for a home milling and mixing plant. Home mixing demands expertise in buying straight foods and in formulation to ensure a palatable and balanced food. The ultimate choice between the two systems depends on the relative costs of either the purchased or the home-mixed concentrate, and whether labour is used more efficiently on the farm attending to the stock or mixing food. Many farm mixing plants are highly automated and require only a little labour.

The most commonly used mixtures for cows contain about 16 per cent crude protein and at least 10·7 MJ per kg and are given at the rate of 4 kg per 10 kg milk. Alternatively a mixture containing about 18 per cent crude protein and 12·3 MJ per kg can be given at 3·5 kg per 10 kg milk. Mixtures of even higher concentration which can be given at 3·2 kg per 10 kg milk are feasible, but will have to include fats. Six examples of concentrate mixtures are given in Table 7.2. Each one contains a variety of ingredients which should increase the acceptability of the mixture. Balanced rations are theoretically possible from a mixture of only two ingredients, but in most dairy herds a few cows will invariably reject one specific ingredient. Palatability can also be increased by adding molasses and locust beans. A dusty mix can be unpalatable, and can be improved by adding a small quantity of water to the mixture.

When one ingredient is used as a substitute for another, factors such as relative cost, palatability and possible toxicity should be considered. A few feeds balanced for milk production, such as bran, palm-kernel cake, coconut cake and maize-gluten feed, can be added to a mixture up to a maximum of 15 per cent of the total without affecting the feeding rate. These foods should be regarded as additions to the mixture rather than replacements.

A farm mixture should be thoroughly mixed, and to achieve this the smaller amounts of ingredients such as the minerals should be added together first, mixed with about 10 per cent of the total amount, and then added to the bulk of the other ingredients. This technique is important with all mixtures, but in particular with those incorporating urea. Care must be taken during storage to ensure that layers of ingredients do not separate, and it is advisable to agitate the mixture before feeding it to the cattle.

Table 7.2. Examples of the formulation of six concentrate mixtures for dairy cows

Ingredient (% by weight)	Feeding rate (kg per 10 kg milk) 4·0			3·5		
Barley	65·0	60·0	60·0	10·0	15·0	22·5
Wheat	—	—	—	12·5	20·0	11·0
Oats	—	12·5	10·0	—	—	—
Maize	—	—	—	30·0	10·0	30·0
Flaked maize	—	—	—	20·0	20·0	20·0
Dried sugar-beet pulp	12·5	—	—	—	—	—
Wheat middlings	—	—	10·0	—	—	—
Cotton cake†	—	—	10·0	17·5	—	—
Soya bean meal*	20·0	15·0	7·5	7·5	20·0	12·5
Bean meal	—	10·0	—	—	12·5	—
Urea	—	—	—	—	—	1·5
Minerals						
20% Ca: 6% P	2·5	2·5	2·5			
25% Ca: 12% P				2·5	2·5	2·5

*Extracted
†Decorticated
(Adapted from *Feeding concentrates to dairy cows*)

Fats and Oils

Diets which are low in fat, i.e. 1 per cent or less, can cause lowered yields of milk. Fortunately most basal diets given to dairy cows provide an adequate amount of fat, and thus the response in milk yield to additional fat is small. The addition of small amounts of fat to a concentrate is beneficial in increasing palatability and in aiding the process of pelleting. Also, because fats are a rich source of energy, they are an attractive ingredient in rations for high-yielding dairy cows.

If too much fat is added, rumen digestion is disturbed and the apparent digestibility of both carbohydrates and protein is reduced. The development of methods of 'protecting' fat by mixing it with protein feeds, drying and treating with formaldehyde may allow the future use of higher levels of fat than in the past. With this technique, the energy content of the concentrate could be increased, and thus a smaller amount of concentrates would be required per 10 kg milk. At present, 3·5 to 4·0 kg concentrates are usually required to supply the nutrients for 10 kg milk but with fat-enriched concentrates the amount may be lowered to about 3·2 kg.

The addition of dietary soya oil normally causes a slight decrease in the proportion of milk fat, whereas oils rich in palmitic acid tend to increase the content of milk fat. Dietary tallow has little effect on the content of milk fat. Depressions in milk protein content have occurred as a result of adding a wide variety of fats to the diet.

Excessive proportions of unsaturated fatty acids in the diet may cause the milk and body fat to be soft, and in young cattle this excess may induce a deficiency of vitamin E which causes muscular dystrophy.

The incorporation of specific oils and fats into dairy concentrates can predictably alter the fatty-acid composition of milk fat, and hence the physical properties of the fat. These effects may become important in the butter manufacturing industry.

Urea and Biuret

Non-protein nitrogen compounds can be useful sources of nitrogen for ruminants (Fig. 2.2), because of the ability of the rumen micro-organisms to convert such simple substances into their own microbial protein. Urea is the most widely used compound for animal feeding, as it is relatively cheap and has a nitrogen content of 46·6 per cent, which is equivalent to a crude protein content of 291 per cent, i.e. $46·6 \times 6·25$. Urea is a white crystalline solid which, in the rumen, rapidly forms ammonia. An excess of urea can produce a toxic amount of ammonia, which enters the blood, and urea should be given in a way which slows down its rate of breakdown.

In feeding practice, no more than 20 per cent of the total food crude protein of the dairy cow should be supplied as urea, and where possible this should be given in the form of frequent, small intakes. Concentrates should contain a maximum of 15 kg urea per t (Table 7.2). The diet should also contain a source of readily available energy so that microbial protein synthesis is encouraged and wastage of nitrogen reduced. Urea contains no energy, minerals or vitamins, and these must be provided in the diet if urea is a substitute for protein. The proportion of crude protein which can be substituted as urea depends on the M/D concentration of the diet and the degradability of the other food proteins in the rumen, and hence urea is not equally effective under all conditions.

Low-yielding dairy cows can use concentrates containing urea efficiently, as concentrate intake is low. Conversely, high-yielding cows, which may receive a high proportion of their total ration as concentrates, do not make such efficient use of urea. On average, milk yield is about 10 per cent less when urea replaces conventional sources of protein in the rations of high-producing dairy cows. The *ad libitum* feeding of a complete mixed diet would seem to be an effective way of feeding urea. It should always be carefully and intimately mixed in the feed to avoid local concentrations, and hence toxic excesses.

Biuret is produced by heating urea, and contains the equivalent of 255 per cent crude protein. Unlike urea it is not toxic, but is more expensive and is rarely used for livestock at present. Foods which contain non-protein nitrogen must carry a statutory declaration giving the protein equivalent of their non-protein nitrogen.

Blocks and Liquid Feeds

Urea is used widely in solid blocks and supplementary liquid feeds. The blocks weigh 15 to 25 kg and contain a readily available source of energy, including starch, glucose, dried solubles and cereals supplemented with minerals and vitamins. Crude protein contents range from 12 to 35 per cent, and often half of this is derived from urea. Cattle have free access to the blocks, and intake is restricted by either the high salt content of the block or their hardness. As a result the blocks must be licked, and thus the danger of excessive intakes of urea is avoided.

Solutions of urea containing molasses, minerals and vitamins are supplied in special feeders in which the animal licks a ball floating in the liquid, but cannot drink the liquid directly. Intake is restricted by the slow action of licking, but this is no guarantee that every animal in a group has an equal intake.

Both blocks and liquid feeds enable stock to have a small and regular intake of non-protein nitrogen and energy, and this can increase the consumption of low D-value forages. With high-yielding cows receiving concentrates and high D-value forage, there is thus little or no place for blocks and liquid feeds. However, under certain low-intensity conditions where the animals are given rations of low D-value and low protein content, the blocks and liquid feeds are suitable. These specialized foods increase the overall nitrogen content of the diet, and may increase digestibility.

Blocks are only justified economically if they act as true supplements and form only a small part of the ration, e.g. 5 to 10 per cent by weight. When the correct amount of a block is consumed, the advantage lies in its convenience and its capacity to stimulate the intake of low D-value forage, which should be supplied *ad libitum*.

Brewers' Grains

Brewers' grains or 'draff' are a by-product of the brewing industry, and consist of the insoluble barley residue after the liquid 'wort' has been drained away. Ensiled brewers' grains contain about 28 per cent dry matter, containing 10·0 MJ per kg dry matter with 15 per cent digestible crude protein in the dry matter. The wet food is often dried to about 90 per cent dry matter, and both the wet and dry products are classed as a concentrate.

The wet grains can be stored for future use in a silo on a firm, well-drained site by ensuring that air and water are completely excluded. The sides of the silo should be airtight, and the top must be covered with polythene sheeting which is uniformly weighed down with tyres.

Brewers' grains are highly palatable and can be offered in amounts up to 16 kg per cow daily to supply both energy and crude protein. They are, however, low in soluble minerals, sodium and potassium, and also in calcium. A specifically formulated mineral supplement high in these minerals and magnesium should be added at a rate of 7·5 kg per t wet grains.

Distillers' grains from whisky distilleries are similar to brewers' grains, but can have a more variable composition, including a higher protein content, depending on the type of grain used. The distillers' solubles are often mixed with the grains and dried to produce 'dark grains'. An undried mixture of malt distillers' grains and pot ale syrup containing about 40 per cent dry matter and 12.2 MJ per kg dry matter is also available. This product can be given in amounts up to 10 kg per cow daily, and may replace a balanced concentrate on a dry matter basis.

Further Reading

Briggs, M. H. (Ed.), *Urea as a protein supplement*, 1967, Pergamon Press, London and New York

Feeding concentrates to dairy cows, ADAS Advisory Leaflet No. 525, 1975, Ministry of Agriculture, Fisheries and Food, London

Forage maize, production and utilization (eds. Bunting, E. S., Pain, B. F., Phipps, R. H., Wilkinson, J. M. and Gunn, R. E.), 1978, ARC, Great Portland Street, London

Frame, J., 'The potential of tetraploid red clover and its role in the United Kingdom', *Journal of the British Grassland Society*, 31, 1976, p. 139

Green fodder crops, Fodder root crops, Forage maize, Cereals and *Field beans*, Farmers' Leaflets Nos. 2, 6, 7, 8 and 15, 1983, National Institute of Agricultural Botany, Huntingdon Road, Cambridge

McDonald, P., Edwards, R. A. and Greenhalgh, J. F. D., *Animal Nutrition*, 3rd ed., 1981, Longman, London

Turnips and swedes—production, harvesting and utilization, Bulletin No. 9, 1975, North of Scotland Agricultural College, Aberdeen

White clover, Publication No. 99, 1983, Scottish Agricultural Colleges, Auchincruive, Ayr

CHAPTER EIGHT

Winter Feeding Systems

Planning the Feed Supply—Distributing the Feed—Hay Feeding—Silage Self-feeding—Forage Boxes—Other Silage Systems—Concentrate Feeding—Kale Feeding—Complete Feeding—Water Requirements

When the level of nutrition for a herd of cows has been decided, and the types of food which are to be grown or purchased to meet their needs, it is then necessary to ensure that a sufficient supply of the foods is available and that this reaches the cows in an efficient manner.

Planning the Feed Supply

In order to ensure a sufficient supply of feed for a herd during the winter, it is necessary to estimate the probable length of the winter feeding period, and to know the number of cows in the herd and the amount of each type of food which they will consume each day.

The intake of an average cow of one of the larger breeds is limited by appetite to about 16 kg of dry matter per day. For a 180-day winter this totals nearly 3 t of dry matter, which includes both concentrates and forage. The ratio of concentrates to forage will vary according to the feeding policy on the farm but if, for example, 1·25 t is in the form of concentrates, then approximately 1·75 t of forage dry matter will be required per cow. At typical levels of dry-matter content, this would represent a little over 2 t of hay or 7 to 8 t of silage per cow for the winter.

To take a specific example, if the full winter feeding period can be expected (allowing a generous safety margin) to extend from 15 October to 15 April, and 80 cows are to be allowed 35 kg of silage per head per day, the amount of silage required for the herd will be $182 \times 80 \times 35$ kg = 510 t. If silage is also fed to young stock an appropriate addition would be made. If it is clear that it will not be possible to supply enough forage for the herd from the farm, arrangements should be made to purchase supplementary food early, before the price rises in the winter.

The herd should be introduced to winter rations gradually; grazed grass is their natural food and if the change to conserved foods is made suddenly milk yields will suffer. The degree to which use can be made of autumn grass will vary widely with the climate and soil type, but in general it is wise to start supplementing grass with other forage before it appears to be necessary; the exact date will vary from year to year, but mid-September is an average time for most parts of the country. The herd will normally be housed at night for some time before being housed all day.

Turning out to grass in spring should be an equally gradual process; cows romp about like lambs when first turned out, and it is good practice where possible to put them first into an old pasture to avoid damaging a more valuable sward. Grazing for 1 hour will be enough on the first day, and concentrate and forage feeding can be gradually reduced as the period of grazing lengthens. The date of turning out is as variable as that of autumn housing, but a national average probably lies somewhere in the first half of April.

Distributing the Feed

The objective in distributing feed to the cows is to ensure that the correct quantity is given to them at the correct time, in a way which makes the minimum demands on both labour and capital expenditure and which allows (so far as modern housing systems permit) each cow to have her fair share.

Cows may have to forage for their bulk feeds for themselves, as in the grazing of kale and other forage crops or in the self-feeding of silage, or they may have the feed brought to them. Delivering the feed may involve nothing more than throwing sugar beet tops from a trailer on to a field; at the other end of the scale it may involve a complex process which mechanically extracts high-dry-matter silage from a tower silo, mixes it with weighed amounts of concentrates and distributes it, still automatically, into mangers. However, most bulk feeds are probably put into mangers either by hand or from a forage box. These mangers may be accessible to machines, as in most cubicle layouts, or may be reached only by hand barrow, as in many cowsheds. In this chapter, it is assumed that a manger is available and that it can be served by trailer or forage box.

To prevent the cows from stepping into the manger there will be some form of feed barrier which, if properly designed, should also help to reduce waste of food and bullying. These barriers vary greatly in elaboration and cost; the simplest form is a kerb (sometimes a railway sleeper on edge) with a single rail or wire over it, but in most circumstances the cost of a rather more elaborate design can be justified, in which the animals have to put their heads between vertical or diagonal divisions. Examples of this type of barrier include tombstone and diagonal barriers (Fig. 8.1). Each Friesian cow should be allowed 600 to 700 mm of barrier. The most refined and expensive form of barrier is the yoke, usually self-locking, in which each cow is secured by the neck until released; yokes have a place, but only for special purposes.

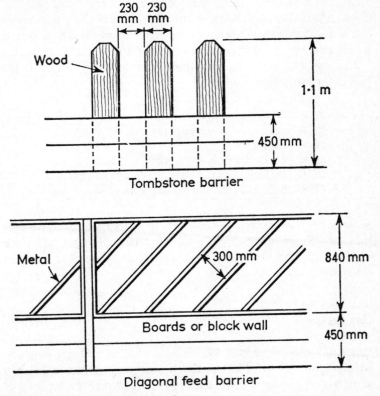

Fig. 8.1. Two types of feed barrier

Judging the quantity of a bulk feed by eye is not easy, but it is important that the correct quantities are given, to ensure both that the cows receive the correct amount of nutrients and that the supply of bulk feed is not used up too quickly. Weigh cells can be fitted to forage boxes and to foreloaders, but where such sophisticated aids are not available, a trailer or forage box full of feed should where possible be weighed on a weighbridge periodically. It is an important part of management to see that the supply of forage lasts to the end of the winter; a change of feed in late winter can depress milk yield severely. If silage, for example, seems likely to run out, it is better to reduce the quantity offered daily in good time, and to make up the deficiency with some other food. This practice ensures that the cows receive at least some silage each day throughout the winter.

Unless bulk feeds are given *ad libitum*, they are usually distributed twice a day at times convenient to the staff. Cows are creatures of habit, and the times of distribution should be as regular as possible. Foods which may cause taints in the milk should not be given shortly before milking, and it is better not to offer feed in the 2-hour period before milking in order to avoid a reduction of concentrate intake.

Feeding schedules will vary from farm to farm, but a simple programme would be as follows:

Time
05.30   Milking—Concentrates (maximum 5·5 kg)
08.00   Bulk feed
12.00   Concentrates in the mangers
15.00   Milking—Concentrates (maximum 5·5 kg)
17.00   Bulk feed (distributed during milking)

Whatever the schedule, it is important to alternate feeds of concentrates, which are low in crude fibre, with feeds of forages, which are high in fibre. Hay in the long form is particularly useful as an aid to rumination.

Hay Feeding

Hay is most usually handled in bales of conventional size weighing an average of 23 kg (Chapter Five). Distributing these bales into mangers has not been successfully mechanized, but the layering of the hay within the bales makes them comparatively easy to divide and distribute by hand. Such bales are particularly suitable for use in cowsheds and other buildings with limited access. It is easy to keep a check on the quantity of hay being given, since the bales of any particular crop are usually of fairly constant weight.

In the past few years, big bales have become increasingly popular, largely owing to the greater ease of handling the bales from the field. These bales may be either cylindrical or square, and weigh from 300 to 600 kg. They are too heavy to manhandle and are thus difficult to distribute in many existing buildings. However, efficient specialist handling equipment is available for these bales, and where such equipment is used and building layout is suitable, distribution of hay can become much less laborious.

Silage Self-feeding

This system, which first became popular in the middle 1950s, is simple in that the cows help themselves, grazing the face of the silo. The system requires fairly exact dimensions in the silo: the settled height of the silage should not exceed about 2·3 m for Friesians or they will be unable to reach all of it, and the width of the feeding face should be about 200 mm per cow. This width can be reduced slightly, provided the cows have 24-hour access to the silo and are not being severely restricted in intake. It is unwise to exceed the width of 200 mm by much, or the silage will not be eaten quickly enough to prevent it from being spoiled by exposure to the air. The cows should be held back from the silage by some form of barrier. Solid barriers such as tombstones are difficult to move and rejected silage builds up behind them; the most usual type of barrier is either an electrified wire or an electrified pipe (Fig. 8.2) suspended with plastic baling string from long angle irons driven into the face of the silage. The electric pulse is brought to the pipe by an insulated wire from an electric fencing wire running along the top of the silo wall. The pipe or wire should be approximately 850 mm above the floor for Friesian cows, and there should be some means of reducing the shock delivered by the fencer.

With an efficient barrier, consumption can be controlled by the distance which the barrier is moved each day; silage intake can be restricted to as little as 13 to 15 kg per head per day, provided that the cows have 24-hour access to the silo, which should be lit at night, and provided that other foods are given in the mangers in addition.

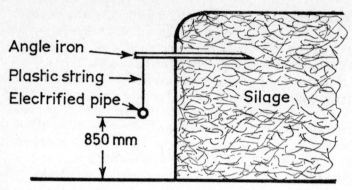

Fig. 8.2. Electrified barrier for self-feed silage

A criticism made of self-feeding is that it leads to overeating by the cows, but this can be overcome simply either by restricting intake with the barrier or by offering alternative feeds. Another disadvantage is stated to be the high degree of wastage, but with well made silage and good management this should not be serious. Probably the most common reason for giving up self-feeding is that herds tend to expand and it is often difficult or impossible to increase the width of the silo face proportionately.

Even where silage is self-fed to appetite there should be facilities for manger feeding, even if very simple ones, so that concentrates and alternative forages can be given when required.

Forage Boxes

Forage boxes are trailers with high sides and usually a moving floor which brings the contents against a set of beaters at the front or rear, which tease out the material on to a cross-conveyor which throws it out at one side. By selecting a suitable power-take-off speed and tractor gear the driver can deliver the desired amount of silage or other feed into the mangers. Such boxes have an average capacity of about 7·5 m³ of silage and are filled from clamp silos, by foreloaders, grabs or specialist silage-unloading machines; from towers, they are filled by top or bottom unloaders. Different makes of forage box vary in their ability to handle material of different chop length, but all types require at least some degree of chopping or laceration in the silage. Some forage boxes can be fitted with concentrate dispensers which mix measured amounts of concentrates with the forage as it is discharged. Forage boxes allow a measured amount of silage to be put out, and they permit greater flexibility of layout than self-feeding, for which the housing and the silo need to be close together. Forage boxes can be used for feeding more than one herd if the distances are not too great, and young stock and beef animals in addition if the housing layout is suitable. Forage boxes are, however, large and cumbersome machines which require careful driving and generous passage widths, with a distance of 9 m in which to turn through 90 degrees. They are also susceptible to mechanical breakdown, and some alternative method of feeding has to be available for emergencies.

Other Silage Systems

In a fully mechanized system the silage is delivered along the mangers by fixed conveyors, which may be of the auger, chain-and-flight, tapered-bed, belt-and-brush or other types. They usually form part of a fully mechanized system including tower silos with mechanical unloaders. Despite its obvious attractions in saving labour and in permitting frequent feeding of both silage and concentrates, this system has not gained widespread popularity in this country. The reasons for this include the high capital and maintenance costs, and the inconvenience caused by mechanical breakdowns. There is also a lack of flexibility of layout, since all the mangers have to be reasonably close to the silos, and the difficulty of determining exactly how much silage is being given, on account of the variable and unpredictable rate of performance of the silo unloaders.

Intermediate between self-feeding and full mechanization are the many systems which have been evolved by ingenious farmers. A typical example is the use of a foreloader or buckrake to move large heaps of silage into feeding passages, which are usually just wide enough to admit the tractor and in which a certain amount of hand forking is required. A refinement of this system is the use of a silage cutter, mounted on either the three-point linkage or a foreloader, which cuts out a clean block of silage weighing about 500 kg. These blocks can be set along the feeding face and forked to the cows as required. Since the silage in the block is practically undisturbed, it is slow to deteriorate. For the same reason the silage remaining in the clamp keeps better because little air is admitted into it.

In another system, not now much seen, a long face of the silo is exposed, allowing some 600 mm of face per cow, and rationed amounts of silage are cut out by hand and thrown down behind the barrier. The advantage of this system is that it encourages a high intake of silage, but its great disadvantage is the large surface area of silage which is exposed to the air and hence deteriorates.

Concentrate Feeding

The average cow of one of the larger breeds, with a liveweight of about 600 kg, can eat no more than about 5 to 5·5 kg of concentrates during each of its two daily visits to the milking parlour, i.e. 10 to 11 kg per day. Where an effort is being made to achieve high peak yields, it may be desirable to offer more than this quantity, and then it is common practice to give a third and sometimes a fourth feed of concentrates outside the parlour. However, recent ideas on nutrition suggest that too much emphasis was placed in the past on achieving a high peak yield, and when good forage is given *ad libitum*, the distribution of concentrates over the lactation is not so important. The total weight of concentrates consumed per lactation has still a dominant effect on lactation yield.

Nearly all cows are given concentrates during milking, whether in cowsheds or parlours. Reasons advanced for not feeding in the parlour are that it speeds up the milking routine and causes less upset to the cows, as well as avoiding the considerable cost of feed storage and dispensers. However, milking time is the only opportunity of treating a cow as an individual which most modern systems permit, and most farmers take this opportunity to feed the cows in the parlour according to yield. Except in the comparatively few herds which are large enough to justify division into three or more groups, feeding in the parlour according to yield seems likely to remain normal practice. Even in large herds, close grouping is not the usual practice when the cows are at grass, and it is desirable to have some way of feeding individual cows which calve during the summer. The performance of concentrate dispensers in parlours must be checked regularly, and in particular every time the type of food is changed.

Most concentrates are now delivered either in bulk or in easy to handle 25-kg bags which facilitate control of the quantity being given. Where the mangers are not used for bulk feeds (e.g. in self-feeding herds), the concentrates can be put out at times convenient to the staff and the cows let through to feed when required—perhaps when the late-night inspection is made. Where the feeding area is shut off by a gate, this may be opened by a time switch.

Where larger quantities of concentrates are being handled, a tractor-powered concentrate dispenser may be justified. These machines, which may be either trailed or mounted on the three-point linkage, consist of a hopper with a capacity of 750 kg to 2 t, which can be filled either from bags or direct from a bulk bin. The concentrates are delivered to the mangers by side-discharging auger. Owing to the comparatively good flow characteristics and predictable physical properties of concentrates, these dispensers can be much simpler and cheaper machines than forage boxes.

Out-of-parlour concentrate dispensers which are programmable offer full automation of concentrate feeding, with complete control of each individual cow's intake. Every cow which is to receive concentrates, usually in practice the whole herd, wears a plastic collar which carries a transponder. This transmits an electronic signal which identifies the cow when she enters one of the feed stations and her identity is relayed to the control unit in the farm office which, having been programmed, releases the correct amount of feed to the cow. The programme ensures that each cow can normally eat only a quarter of her daily ration in any period of 6 hours, and any balance not consumed is carried over to the next 6-hour period but not beyond the 24-hour period. The control unit has a keyboard for entering the cow numbers and rations, and either a visual display or a printer which states the total weight of feed programmed and the total consumed during the past 24 hours; it also lists animals which have not eaten their full ration. It is usual to provide one feed station for every 25 to 30 cows.

Out-of-parlour feeders aim to give each cow an exact ration and to ensure that intake of concentrates is spread evenly over the 24 hours, which may have advantages compared with offering concentrates only twice a day in the parlour. These advantages are claimed to be higher fat contents in the milk, better utilization of the concentrates and less digestive disturbances, although these latter advantages can to some degree be obtained by giving a third feed of concentrates in a manger. Out-of-parlour feeders remove most of the labour from concentrate feeding. They also remove the need for in-parlour feeders and can, for example where silage is self-fed, remove the need for a separate feeding area. They also reduce the need to group cows.

Kale Feeding

Kale is by far the most common of the grazed forage crops, although rape, stubble turnips and other crops are not uncommon. Kale is most usually strip-grazed, and an electric fence in good working order is required. If the wire is placed in front of the kale, the animals will graze the crop and not trample it. About 3 m of kale frontage is required per cow, and it is wise to select a field which is the correct length for the size of herd. If possible, the field should be adjacent to a grass field in which the herd can rest when not grazing the kale.

With a kale crop sown on an old grass sward, poaching is much less severe than on a field which has been ploughed and cultivated, and the cows will not be as dirty. Mud on legs and udders is unfortunately a problem with kale grazing, and this can cause sore teats and extra work at milking time. In frosty weather, the cows can develop sore feet, and it is also unwise to feed kale which has been substantially damaged by frost. Where possible, it is safer and more economic to complete the kale grazing before the arrival of severe frosts, i.e. in the period October to December. Another way of avoiding the mud problem is to cut the kale with a forage harvester and distribute it with a forage box.

Approximately 1 ha of kale is required for feeding 20 to 25 cows. Under good conditions, a cow will eat about 20 to 25 kg in a 3-hour period in the morning, and this amount will contain 3 to 4 kg dry matter. The kale dry matter can replace a similar weight of cereal and silage dry matter in the ration, and also increase the crude protein content of the diet. It is useful to check intake by making some spot weighings.

Complete Feeding

In this system, the forage and concentrates are intimately mixed together and delivered to the cows by a mixer wagon. This is a sophisticated and expensive machine comprising a steel V-shaped body containing three or more cross-augers which mix and discharge the feed to one side. It can also incorporate weigh cells between the body and the chassis which indicate the weight of each ingredient in the mix. The boxes vary in capacity from 4 m³ to about 10 m³. The density of a typical mix of silage and barley averages about 500 kg per m³.

The essence of the system is the thorough mixing, which prevents the cows from picking out one particular food. This means that a feed containing concentrates can be given *ad libitum* without danger of the cows overeating the concentrates and neglecting the forage. This ability to feed *ad libitum* means that cows in peak lactation can be given the opportunity of realizing their full potential. However, feeding at this level can only be justified for cows in peak lactation, and the herd must be divided into three or more groups according to calving date. Alternatively the complete feed may provide a balanced ration up to a certain level of nutrient intake with additional concentrates being given in the parlour.

Where close grouping is practicable, it is possible to design a complete diet for each group which is closely matched to its needs, and also to ensure that maximum use is made of home-grown forage, particularly in later lactation. The use of weigh cells can also ensure that the correct quantities are given.

To suit a mixer wagon, silage must have a high dry-matter content and be finely chopped; this means that complete feeding is essentially part of a highly mechanized system and thus best suited to larger herds.

A further advantage of the preclusion of selection by the cows is that the foods need be distributed only once a day instead of twice or three times. However, this advantage must be weighed against the fact that filling a mixer wagon and distributing the feed takes considerable time and skill.

Mixer wagons are ideal for making use of by-products and other foods which may not be sufficiently palatable if offered alone. It is possible to use a forage box for complete feeding, the foods being placed in layers and the action of the beaters and discharge auger having some mixing effect. The effectiveness of this depends on the type of foods involved, but cannot compare with that of the mixer wagon.

In order both to maximize intake and to avoid digestive upsets, it is important that the cows consume the complete ration in small, frequent feeds, and that they have 24-hour access to it. If for any reason the cows are left without food for more than 2 hours, a food of low energy content must be given before the complete diet is reintroduced.

Where moist rations have been well mixed and compressed so that they contain little air, secondary oxidation is restricted and in some circumstances it has been found possible to distribute the feed on alternate days. This compression of the feed appears also to be partly responsible for the increased dry-matter intake which is commonly found among cows, particularly those with higher yields, on complete diets. This increase has been estimated at 10 per cent or more.

Foods used in complete diets include silage, swedes, brewers' grains, cereals, maize, high-protein concentrates, molasses and chopped hay and straw. The exact formulation of the ration depends on the availability of the foods on the farm, on the relative costs and the desired ME content.

A complete diet should have a crude protein content of 12 to 14 per cent and a crude fibre content of 16 to 20 per cent in the dry matter. Most diets should also contain a minimum of 40 per cent forage and 40 per cent concentrates on a dry-matter basis, including minerals. For cows in the first 100 days of lactation the complete diet should have an ME of 12·0, in days 100 to 200 an ME of 10·0 to 10·5, and after 200 days an ME of 8·5.

Water Requirements

The water requirements of cattle are supplied from water formed during various metabolic processes within the body, from the water in the food, and from the water drunk voluntarily. The sum of the latter two sources is usually defined as the water requirement, and is 3·5 to 4·1 kg water per 1 kg food dry matter for non-milking animals at an environmental temperature of 10 to 21 °C. If cows are in milk, an extra allowance of approximately 1 kg water per 1 kg milk produced is required. Water intake increases as the environmental temperature increases, but this factor is of little consequence under the temperate conditions which prevail during the winter in Britain.

The daily milk yield of the cow and the dry-matter content of the animal's diet are the major items affecting the amount of water drunk, and from Fig. 8.3 it is possible to determine the drinking-water requirements of Friesian cows. As a safety measure, an extra 10 kg water per cow should be added to the values determined from Fig. 8.3. On average, a maximum daily supply of 80 kg drinking water per cow would suffice on 95 per cent of occasions in most herds.

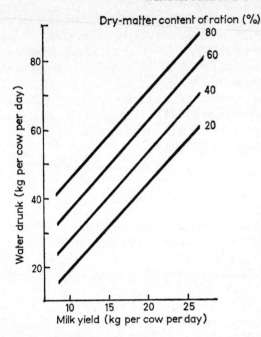

Fig. 8.3. Relationship between the daily intake of drinking water and the daily milk yield per cow on rations with four different dry-matter contents (From *Animal Production*)

To ensure that milking animals are not short of drinking water, it is necessary to know both the average water requirement per day and also the period of maximum water demand. This is usually in the 6-hour period between 15.00 and 21.00, when 40 per cent of the total daily intake may be drunk. The peak demand for water occurs 1 to 3 hours after the evening milking, and the flow rate of water to the troughs and bowls must be adequate at this time. Water-supply pipes and fittings with a 19 mm bore are advisable. One water trough 1·8 m long is required for 50 cows on a ration of low dry-matter content or for 30 cows on a ration of high dry-matter content. The optimum height of the top of the water trough is about 900 mm, and a guard rail about 150 mm from the trough will prevent cows from dunging in the water. Troughs should be kept clean, since fouled water can quickly deter cows from drinking.

Water troughs should not be sited in either strawed areas or in confined parts of the building such as blind passages. The position of drinking points should allow some cows to drink while others have space to walk past them without jostling. Good sites for troughs are in the feed area, or at the end of a row of cubicles, or near the exit point of the milking parlour, provided that these positions do not lead to congestion, and hence to some cows being short of water. Cows can drink rapidly, 16 to 27 kg water per minute, and water troughs with a large capacity are preferable to bowls in loose housing. The troughs should be sited so that 10 per cent of the herd can drink in comfort at any one time, with 40 to 60 mm of trough face for every cow in the herd.

Further Reading

Castle, M. E. and Thomas, Thesca P., 'The water intake of British Friesian cows on rations containing various forages', *Animal Production*, 20, 1975, p. 181.

Owen, J., *Complete diets for cattle and sheep*, 1979, Farming Press, Ipswich

Raymond, W. F., Shepperson, G. and Waltham, R., *Forage conservation and feeding*, 1978, 3rd ed. (revised), Farming Press Ltd, Ipswich, Suffolk

The mechanization and automation of cattle production, Occasional Publication No. 2, 1980, British Society of Animal Production, Milk Marketing Board, Thames Ditton, Surrey

Milk Composition

Milk is an important source of protein, minerals and vitamins in the human diet, and there has been an increasing interest in its composition in the last 20 to 30 years. Prior to this period, attention had been focused mainly on the fat content, a good 'cream line' being considered a measure of milk quality. However, the solids-not-fat (SNF) content of milk is now regarded as of equal, if not of greater value, than the fat content. Chemical composition determines the ultimate nutritional value of the milk for the consumer, and also has a direct effect on the output of dairy products in the creamery. For example, if the SNF content of milk is reduced by only 0·1 per cent, a batch of 50,000 l of milk will produce approximately 52 kg less milk powder. Now that the importance of milk composition is appreciated, producers of milk are paid on the basis of both quantity and quality, and for most supplies there is a direct relationship between the price received and the weight of milk solids sold.

Milk Constituents

The main constituents of milk are indicated diagrammatically in Fig. 9.1. Virtually all the major constituents are synthesized in the udder from various precursors which are absorbed selectively from the blood (Chapter Two). The cow's diet is the ultimate source of most of the materials used in milk synthesis, and alterations in the amount and type of food affect both milk yield and composition. The products of digestion which are absorbed from the rumen and small intestine into the bloodstream are controlled not only by the ration of the cow but by the types of micro-organism which become established in the rumen. The relationship between diet and the constituents in milk is therefore complex.

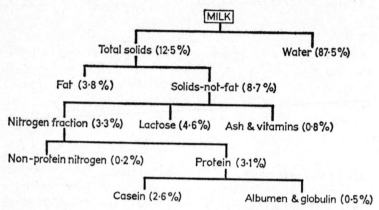

Fig. 9.1. Content by weight of the main constituents of milk in an 'average' sample

Water is the main constituent of milk (Fig. 9.1) and is secreted in association with water-soluble constituents of which the most important quantitatively are lactose, sodium, potassium and chlorine. The amount of water secreted, and therefore the yield of milk, is related closely to the amount of lactose synthesized and secreted in the udder. Milk fat is a mixture of triglycerides containing both unsaturated and saturated fatty acids, and is derived from the fatty acids of the blood triglycerides or synthesized from acetate and β-hydroxybutyrate of the blood plasma. Lactose, termed milk sugar, is the only carbohydrate in milk and is produced mainly from the glucose of the blood. The protein of milk, which is mainly in the form of casein, is derived primarily from the amino acids of the blood (Chapter Two, Table 2.2). The ash in milk contains the major elements calcium, phosphorus, and magnesium, in addition to potassium, sodium and chlorine and a wide range of trace elements including iron, manganese and iodine. Finally, there are considerable amounts of vitamins A and B in milk, with smaller quantities of vitamins C, D, E and K. Vitamins are not synthesized in the udder, but are absorbed from the blood.

Effect of Breed

The composition of milk varies with a number of non-nutritional factors, one of the most marked being the effect of breed. The fat and crude protein content of the milk from recorded cows of different breeds in England and Wales are shown in Table 1.7, and it can be seen that both these two constituents are related inversely to the lactation yield. Friesians, the breed with the highest yield of milk, have the lowest fat and protein contents, whereas the low-yielding Jerseys have the highest contents. Lactose values tend to be slightly higher in the milk from breeds with a high fat content compared with breeds of low fat content and thus the SNF contents (protein + lactose + ash) are also higher. Because of this fairly close relationship between fat and SNF contents, the selection of cows producing a high fat content should also ensure a high SNF content. Selection for a specific characteristic is clearly more effective if it is based directly on a measurement of that factor, and there is a current interest in breeding for an increased protein content, which can only be done by regular protein analyses of the milk from individual animals.

The strain and the individuality of cows within a breed also have a large effect on milk composition. Although an average value from many thousands of analyses may show that the Ayrshire breed has a higher fat and SNF content than the Friesian breed, there is a considerable overlap in values from individual herds of the two breeds. The poorer milk quality from a specific herd of one breed compared with another herd of a different breed can be due to individual breeding, feeding and management. Healthy animals which are known to produce milk with a low fat and SNF content should be culled from the herd, and their progeny either not retained or kept under close scrutiny. In practice, one or two cows giving milk of abnormal composition have little effect on the composition of the bulk milk from a herd of 60 to 80 cows, and thus there is little benefit to be gained from keeping a single Jersey cow in a large herd of Friesians. It is always preferable to select herd replacements from cows yielding milk of high fat and SNF content, since the heritability of both fat and SNF content characteristics is comparatively high.

Age of Cow

As cows become older, the contents of fat, SNF, crude protein and lactose in the milk all decrease slightly. In survey work with Ayrshire cows, the fat content was relatively constant for the first four lactations and then decreased regularly to the eleventh lactation, whereas the contents of crude protein and lactose both declined from the first to the eleventh lactation. With Friesian cows, the average yearly decline in fat and crude protein contents were 0·03 and 0·01 per cent from the first to the tenth lactation. In a commercial situation, only a small proportion of the dairy herd is likely to have had more than four lactations, and age is not therefore a major factor influencing milk composition.

Stage of Lactation and Season

The fat and SNF contents of milk both fall as milk yield increases to a peak at about 2 months after calving. Thereafter, the fat, SNF, and crude protein all increase slowly during the remainder of the lactation. Lactose follows a different pattern, and is highest at the time of peak yield and then declines slowly throughout the lactation, although the size of the fall is more than compensated for by the increases in the other constituents.

Seasonal effects tend to be smaller than lactational effects, but the following changes are generally found. Fat content is highest in October and then falls steadily to a minimum in June. Crude protein contents rise to a peak in May and June, and again in September, with the lowest values in the period from January to March. Lactose contents tend to be high from January to June, but the variations are not large. Total solids and SNF contents are at a minimum in March and April, with SNF at a peak in May and June and total solids at a peak in October.

The effects of season and stage of lactation can combine, and thus cows calving from November to February can be expected to produce milk with low SNF contents. This combination of factors, plus any nutritional deficiencies in winter feeding, can be responsible for the high proportion of milk samples which are low in SNF from January to April.

Type of Food

Seasonal changes in milk composition are linked closely with changes in the amount and type of food offered to the cows but it is stressed that this is a complex relationship. The effect that changes in feeding practice may have on milk composition can often be clearly shown in controlled feeding experiments, but the interpretation and application of these findings in commercial practice may be difficult and even misleading.

One of the most marked changes in milk composition due to diet is the effect on fat content which is termed the 'low-fat syndrome'. Milk fat is highest when the fermentation in the rumen favours the production of acetate, and thus diets which depress this acid will depress the content of milk fat. Such rations contain insufficient roughage, and if a cow receives only 1 to 2 kg of hay and straw per day the content of fat can fall by 1 to 2 per cent. Individual cows may even produce milk with only 1 per cent of fat. The fat content is lowered either if the diet contains a large amount of a highly digestible concentrate such as flaked maize or if the roughage is finely ground before pelleting and feeding. Thus, to avoid low fat contents on winter rations it is wise to feed a minimum of either 3 kg hay or 20 kg silage per cow each day, and to include dried beet pulp in the ration, all of which favour acetate production.

Of equal or greater commercial significance is the seasonal depression in milk-fat content which occurs in April and May when cows are grazing leafy herbage low in crude fibre. The depression can be made even worse if the cows are offered supplementary concentrates with a high energy and low crude fibre content. Thus to guard against a major depression in fat content in spring it is advisable to change slowly from winter to summer feeding over a period of 10 to 14 days, and to reduce the level of concentrate feeding more rapidly than that of the hay and silage. In addition, it is worthwhile to strip-graze the herd tightly behind an electric fence so that the cows graze both the tips of the grass, which are low in fibre, and some way down the stems, which are higher in fibre. In practice it is not always easy to encourage cows to eat hay at this time of year, and it is therefore good practice to ensure that the grass is not too short, and contains a proportion of stem. Grass at this period of the year has a D-value of 75 to 80 and milk yield will not be depressed if the cows are compelled to eat it well down towards ground level.

In the autumn a similar fall in fat content can occur if large amounts of kale and concentrates are given with only a small amount of hay and straw. Again it is essential to increase the ration of hay and to attempt to make the cows eat both the leaf and stem of the kale.

Effect of Energy and Protein

In practice, the most dramatic increase in the protein and SNF content of milk occurs with the change from the winter ration to spring grazing. Increases of 0·4 per cent in SNF content are not uncommon, and indicate the effect of increasing the total nutrient intake of the cows. Young grass has a high feeding value and supplies the cow with an increased level of both ME and of crude protein. In general, it appears that ME intake is the main factor controlling the protein and SNF level in milk; the intake of protein can be as low as 80 per cent of the normal standard before the SNF content is affected. If protein intake is only 60 per cent of the standard, the milk protein and SNF content is, however, reduced. There is often a reduced intake of ME in the diet at the end of the winter feeding period before grass is available, and every effort should be made to provide forages of high D-value at this crucial period. Although the level of protein in the ration may not directly increase SNF contents, it can have an indirect effect by increasing the voluntary intake of the forages and hence the overall intake of all nutrients.

Body condition reflects the cumulative effects of previous feeding, and thus the immediate effects of a poor ration on milk composition can be reduced if cows are in good condition, and body tissues can be mobilized. There are thus excellent reasons for giving extra food before calving to ensure that the animal is in a fit condition to achieve a high peak yield and produce milk with a high SNF and protein content.

The quality of forage, i.e. the hay and silage in the ration, has a dominant effect on the level of SNF and protein in milk. After a poor summer when the quality of forage is low, SNF values tend to be low in the following winter, whereas after a good summer when hay and silage are of relatively high quality, SNF values are higher. Concentrate feeding can influence milk quality, but it is forage quality which is the major factor. In the short term it is often uneconomic to feed extra concentrates to improve the SNF content of milk if forage quality is low and the condition of the cows is unsatisfactory. Feeding and milk composition must be viewed in the long term, probably over the entire lifetime of the cow, and not in short periods of a few months. With this long view, the aim should be to supply the cow with adequate forage of high quality supplemented with concentrates as required to ensure that no major underfeeding of ME and protein occurs.

Milking Intervals

Cows are rarely milked at intervals of exactly 12 hours, and after the longer night interval the fat content in the morning's milk is lower than that in the previous evening's milk. With milking intervals of 15 and 9 hours, the difference in fat content can be 1·0 per cent; whereas with intervals of 13 and 11 hours, the difference is 0·3 per cent. When the total daily milk production is mixed thoroughly in the bulk tank, the difference between night and morning milks is of little commercial consequence apart from a few minor exceptions. For example, to assist in increasing the fat content of bulk milk it may be worthwhile to use morning milk for calf feeding and domestic use. The contents of SNF in morning and evening milk are similar, especially if calculated on a fat-free basis.

The first milk taken from an udder is low in fat, and the content increases as milking progresses, with the highest value in the strippings, i.e. the last milk removed. Thus cows which are incompletely milked will give short-term fluctuations in fat content. Normally this is of no real significance except when milk samples are taken for chemical analysis.

Disease

Mastitis in both the clinical and sub-clinical form can drastically reduce the yield of milk, and in addition decrease the SNF content (Chapter Seventeen). This is due mainly to a fall in the lactose content. There is also a fall in the proportion of the casein nitrogen to the total nitrogen in the milk, and a normal value of about 79 per cent, termed the casein number, may fall to 70 per cent and even to 40 per cent in severe outbreaks of the disease. The chloride content of the milk increases in order to compensate for the effect of the low lactose content when mastitis occurs, and this increase can be used as a further means of diagnosing the disease. All milk from mastitis-infected quarters should be excluded from the bulk tank as an obvious short-term measure, but it is also important to reduce the level of mastitis infection in the herd as a whole in order to improve the SNF content in the bulk milk.

Ketosis (Chapter Seventeen) reduces milk yield, depending on the severity of the condition, and also markedly increases the fat content of the milk. The effect of other diseases on milk composition is not so clear, but any illness which raises the body temperature above 39 °C will depress milk yield and SNF content, with a tendency to increase the fat content.

Quality Payment Schemes

For many years the presumptive legal minimum standards for the composition of whole fresh milk were 3·0 per cent fat and 8·5 per cent SNF. Under the EEC regulations only a minimum of 3·0 per cent fat is now required. However, the wholesale price of milk paid to the producer is based on fat, protein, lactose and SNF content. Slightly different schemes are operated by the various marketing boards, but basically the milk from each producer is sampled and tested for quality at least once per week. A basic pool price is paid for milk of a basic composition, and additions to and deductions from this price are made if quality is above or below this composition. In England and Wales payment is based on fat, protein, and lactose contents, in the Scottish Board area on fat and SNF, and in Northern Ireland on total solids. The basic price of the milk, and the variations in the value of the additions and deductions per litre vary in the areas covered by the different marketing boards. For example, in the Scottish scheme there is a maximum addition of 3·5 p per 1 and a maximum deduction of 3p per 1.

Taints and Flavours

Feed taints in milk can be transmitted from foods such as turnips (Chapter Seven) and weed plants via the blood, but are relatively easy to control by reducing the intake of the offending food and by feeding it to the cows well before the time of milking. Taints can also be absorbed from the atmosphere by having strong-smelling foods and items such as fuel oil and disinfectants in the cowshed and dairy. Illness can cause off-flavours in milk: for example mastitis milk has a salty taste, and that from cows with ketosis is sweetish and unpleasant. Bacterial taints can vary from malty to bitter: these are controlled by techniques of clean milk production. Finally, a rancid taste can occur in milk as the result of the enzyme lipase, which affects the fat. This is now a rare problem and can be avoided by ensuring that the dairy equipment is not old and worn, and by avoiding rapid changes in the flow of the milk along pipelines and around corners.

Keeping Quality of Milk

The length of time which milk will keep at a specific temperature before it becomes sour and unusable depends mainly on the number and type of bacteria which it contains. Thus a measure of the storage life of milk can be obtained by estimating the number of bacteria in milk; this is done in a laboratory and is called the Total Bacterial Count (TBC). Briefly, samples of milk from each farm are tested weekly for the TBC, and a monthly average calculated. Slightly different schemes are operated by the different Milk Marketing Boards but all include a system of penalties as the number of bacteria per ml increase. In England and Wales, an average of 100,000 to 250,000 bacteria per ml incurs a penalty of 0·4p per 1 whereas values under 20,000 gain a bonus of 0·2p per l. In Northern Ireland a count of 500,000 or more bacteria per ml results in a price reduction of 1.0p per 1. The bacteria are derived from cows with mastitis, or dirty udders and milking equipment.

Penalty deductions are also imposed if antibiotics and other inhibitory substances are detected in the milk. These residues from the treatment of mastitis can cause problems in cheese and fermented milk manufacture, and milk from a treated cow should not be mixed with the bulk milk until a period specified by the manufacturer has elapsed since the treatment was given. The penalty for the presence of antibiotics varies from the rejection of the entire milk supply to a wide range of price reductions, depending on the schemes in the various countries. To ensure that milk from quarters treated with antibiotic is not sold, all treated cows should be either marked clearly or segregated for milking last.

Milk in the Diet

The average consumption of liquid milk per person in Britain is 2·5 l per week, and virtually all of this is now heat treated before it is sold to the public. The nutritive value of milk can be assessed on its content of individual essential nutrients, but probably it is best judged by the way it complements and supplements many mixed diets. In the diet of a man doing medium work, the intake of 0·5 l of milk per day will contribute approximately 75 per cent of the recommended allowance of calcium, 50 per cent of the vitamin C, 45 per cent of the riboflavin, 20 per cent of the protein and 10 per cent of the energy. For a child aged 2 to 6 years with a similar intake of milk, the percentage contribution of nutrients from the milk will be even higher.

Milk and dairy products make an important contribution to the supply of nutrients for the human diet, but there has been widespread discussion in recent years of the possible dangers of animal fats to human health and longevity. Animal fats, including milk fat, contain a high proportion of saturated fatty acids and little of the essential polyunsaturated fatty acids, and there are claims that a high dietary intake of saturated fat and a shortage of polyunsaturated fatty acids is a predisposing factor in the production of heart disease. This association must not be disregarded, but many other factors complicate this apparently simple relationship, and the full facts are not clearly established.

Further Reading

Factors affecting the yields and contents of milk constituents of commercial importance, 'Bulletin', Document 125, 1980 (eds. J. Moore and J. A. F. Rook), International Dairy Federation, 41 Square Vergote, 1040, Brussels, Belgium

Falconer, I. R. (ed.), *Lactation*, 1971, Butterworths, London

McDonald, P., Edwards, R. A., and Greenhalgh, J. F. D., *Animal Nutrition* (3rd ed.) 1981, Longman, London

Rook, J. A. F., 'Nutritional influences on milk quality', *Principles of Cattle Production*, (ed. H. Swan and W. H. Broster), 1976, Butterworths, London, p. 221

Rook, J. A. F., 'Nutrition of the cow and its effect on milk quantity and quality', *Journal of the Society of Dairy Technology*, 29, (3), 1976, p. 129

1. Closed-circuit calorimeter for determining respiratory exchange and measurement of the energy content of foods (*Chapter 2*).

2. The application of granular fertilizer, especially nitrogen, is important in producing high yields of herbage. A distributor with a spinner mechanism is ideal for rapid application to grassland (*Chapter 3*).

3. Short, leafy grass with no seed heads has a D-value of 70 and over (*Chapter 3*).

4. Tall, stemmy grass with the heads fully emerged has a D-value of only 55 (*Chapter 3*).

5. Equipment for electric fencing includes a tripod, reel and battery unit (*Chapter 4*).

6. Strip-grazing with an electric fence enables the correct amount of herbage to be given to the herd each day (*Chapter 4*).

7. A heavy crop of hay with a D-value of 50 to 55 has little place on the intensive dairy farm (*Chapter 5*).

8. Well-made hay with a minimum D-value of 60 will save concentrates in the winter (*Chapter 5*),

9. Silage towers require specialized equipment for filling and emptying, and herbage of 35 to 40 per cent dry matter (*Chapter 6*).

10. A precision-chop forage harvester cutting herbage to a length of 15 to 20 mm will improve silage fermentation. Note the additive container (*Chapter 6*).

11. Efficient sealing of the clamp silo to exclude air and rain is vital to reduce losses and wastage (*Chapter 6*).

12. Strip-grazing kale in mid-winter with an electric fence is a cheap and convenient way of producing food for the herd in mid-winter (*Chapters 7 and 8*).

13. A self-unloading forage box is ideal for rapidly discharging grass silage into cattle feeding troughs (*Chapter 8*).

14. Complete feeds must be intimately mixed in a mixer wagon to prevent cows from selecting one specific ingredient (*Chapter 8*).

15. This simple feed barrier is effective and cheap (*Chapter 8*).

16. An infra-red milk analyser can analyse hundreds of milk samples per day automatically and print out the results (*Chapter 9*).

17. Milk in refrigerated bulk tanks is kept at 4° C and collected daily (*Chapter 10*).

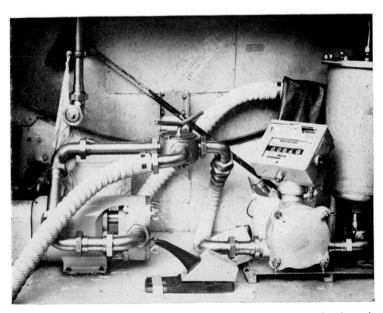

18. The amount of milk pumped from the bulk tank to the tanker lorry is recorded automatically and printed out (*Chapter 10*).

19. Automated cluster removal speeds up the routine in a doubled-up herringbone parlour with low-level jars (*Chapter 11*).

20. The abreast is the cheapest type of rotary parlour and is mechanically simpler than other types (*Chapter 11*).

21. Eye-level jars in a herringbone milking parlour allow easy reading of the graduations (*Chapter 11*).

22. The outline of a new dairy building is softened by planting a few well-placed trees (*Chapter 12*).

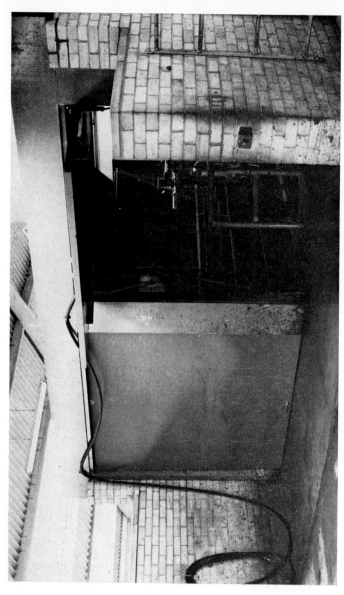

23. An up-and-over door between the milking parlour and the collecting yard can be closed between milkings (*Chapter 11*).

24. Strawyards for dairy cows require far more bedding than cubicles, and may cause problems with mastitis in late winter (*Chapter 12*).

25. Cubicles for heifers can be constructed of wood, but the ideal bed is difficult to achieve (*Chapter 12*).

26. Cubicles with an adjustable headrail and rubber mats are suitable for young stock (*Chapter 12*).

27. Holding pens equipped with a water bowl and a retaining chain are useful for cows to be examined and inseminated (*Chapter 12*).

28. Above-ground slurry store constructed of silo wall sections can hold the slurry produced over an entire winter (*Chapter 13*).

29. Solid mixtures of dung and straw can be moved efficiently by mechanical cleaners, and stored in a dungstead (*Chapter 13*).

30. Gaps between railway sleepers in a slurry store should be about 25 mm to allow seepage of liquid (*Chapter 13*).

31. A straining compartment in an earth-walled slurry compound (*Chapter 13*).

32. Emptying an above-ground store with a tractor and foreloader is done after the removal of the liquid (*Chapter 13*).

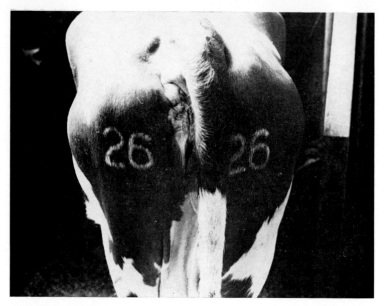

33. Freeze-branding is a permanent method of cow identification for dark-haired cows (*Chapter 14*).

34. Plastic neck collars with numbers on the side and the top can be provided in different colours to aid identification (*Chapter 14*).

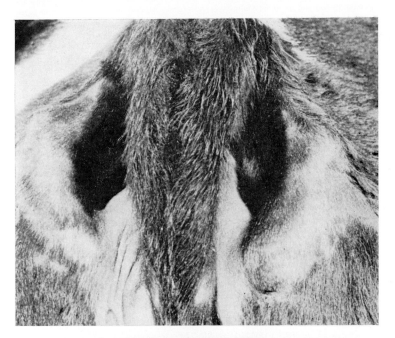

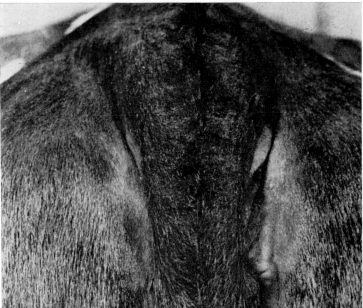

35. Body scoring is a useful aid to cow management: (*above*) score 0, with a deep cavity around the tailhead; (*below*) score 3, with the cavity filled with fatty tissue, and the skin smooth and round (*Chapter 14*).

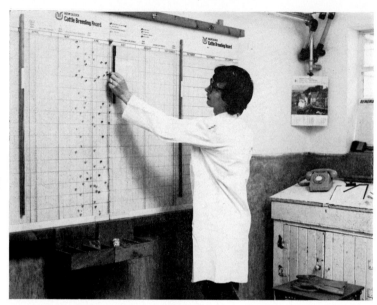

36. A rectangular calendar of the herd breeding records is an important visual aid in the farm office (*Chapter 14*).

37. Steam cleaning of wooden calf pens after each batch of animals will reduce the risk of disease (*Chapter 16*).

38. Young stock in yards should gain 0·6 to 0·7 kg per day on foods of high quality (*Chapter 16*).

39. Individual pens for young calves may have slatted sides and a slatted floor bedded with straw (*Chapter 16*).

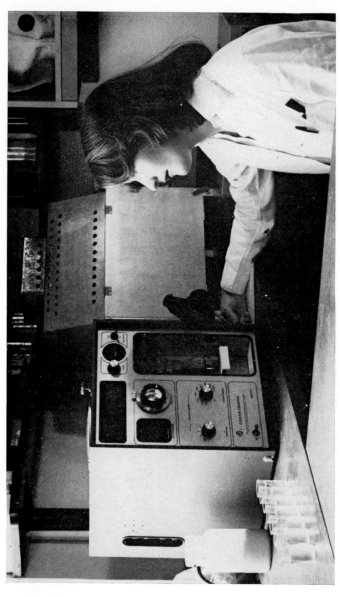

40. The cell count of milk, determined by a Coulter counter, reflects the probable level of mastitis (*Chapter 17*).

41. Cows at pasture will lie down for about 9 hours per day and ruminate for about 7 hours (*Chapter 17*).

42. Walking expends energy, and grazing fields should be close to the farm (*Chapter 17*).

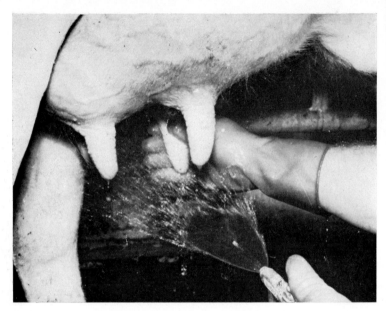

43. Udder washing with warm running water and gloved hands will assist in producing clean milk and reduce the spread of bacteria causing mastitis (*Chapter 17*).

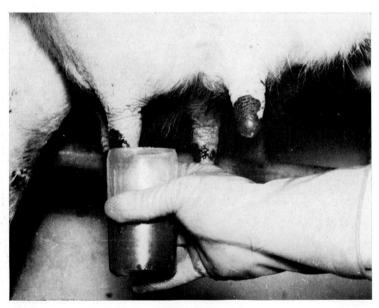

44. Teat dipping immediately after milking is an important part of the mastitis-control hygiene programme (*Chapter 17*).

CHAPTER TEN

Milking and Milking Machines

Udder Anatomy—Lactational Physiology—Rate and Frequency of Milking—Milking Machines—The Cluster Assembly—Conveying the Milk—Producing Clean Milk—Cooling and Storing Milk—Maintenance and Testing—The Machine and Milking Efficiency—Milking Machines and Mastitis

A brief description of the synthesis of milk was given in Chapter Two, but a fuller understanding of the process of milk secretion will be helpful in the planning and operation of an efficient milking routine.

Udder Anatomy

The udder is supported by strong ligaments (Fig. 10.1), of which the most important are the median suspensory ligaments. These are attached to the pelvic bones and to the inner side of each half of the udder, being fused together for additional support. The lateral suspensory ligament (Fig. 10.1) is fastened to the median ligaments at one end and to the body cavity at the other, and forms a supporting sling round the outside of the udder. Weakness of these ligaments can result in dropped and misshapen udders.

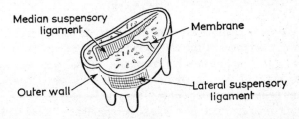

Fig. 10.1. Cross-section of udder, showing the four separate quarters

The udder has four separate compartments. The blood supply and nerve system of the left and right halves of the udder are almost completely independent of each other. The front and hind quarters of each half have a common blood supply, but are separated by a fine membrane and have separate gland and duct systems. Thus all the milk from one teat is produced by the glandular tissue of that specific quarter. In healthy cows, the two front quarters normally produce about 40 per cent of the total milk yield and the two hind quarters about 60 per cent.

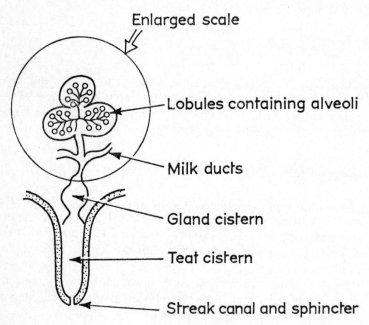

Fig. 10.2.   Cross-section of duct system in one quarter of the udder
(not to scale)

The gross internal structure of the udder is shown in Fig. 10.2. Milk from the lumen of the alveoli is passed via many thousands of small ducts into eight to twelve main milk ducts. These lead into the gland cistern and the teat cistern, and the milk leaves the teat via the streak canal or papillary duct. This canal is surrounded by a ring of smooth muscle fibres known as the sphincter; tightness of this sphincter is among the causes of 'hard milking'.

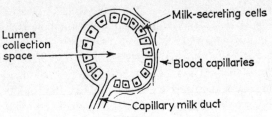

Fig. 10.3. Cross-section of an alveolus

The alveoli are microscopic, balloon-like structures composed of a single layer of epithelial cells (Fig. 10.3). Each cell is in intimate contact with a blood and lymph supply, and the milk is synthesized continuously within the cells and secreted into the lumen of each alveolus and then removed at the time of milking. The alveoli and ducts are surrounded by contractile cells, termed basket cells, which contract when 'let-down' of milk occurs.

Blood, which supplies milk constituents to the udder, comes from the heart via two large arteries each about 10 mm in cross-section. Then, after passing round the alveoli in the udder, the blood enters the venous system and returns to the heart via two main routes. One route is internal, whereas the other is partly external via a vein which passes along the abdominal wall just under the skin and enters the body at a depression termed the milk well. This abdominal vein is known as the milk vein, and its size is often taken as an indication of the cow's potential for milk production; but it is extremely doubtful whether this has any significance.

Lactation Physiology

Briefly, the development of udder tissue is stimulated by a combination of ovarian steroid hormones, i.e. oestrogens and progesterone, other hormones produced by the pituitary gland, and by the developing placenta. At the time of calving, the steroid hormones decrease in amount and lactation is initiated by prolactin from the anterior lobe of the pituitary gland, and other factors. The maintenance of milk production is mainly influenced by pituitary hormones, but other hormones from the adrenals, ovaries and thyroid are also involved.

Milk ejection, or 'let-down', is controlled by the hormone oxytocin, which is secreted by the posterior lobe of the pituitary gland. This hormone causes the alveoli and small ducts to contract, and milk is ejected into the gland and teat cisterns. Milk let-down is mainly a conditioned reflex which is initiated when the cow is subjected to some stimulus which she has been conditioned to associate with milking. Let-down normally occurs within 0·5 minutes of the stimulus, which may be suckling, handling or washing of the teats, the feeding of concentrates or even the sound of the milking machine. It is essential to establish a regular and unchanging milking routine which stimulates let-down with some clear signal and then remove the milk from the cow as soon as possible.

If let-down is incomplete or if milking is delayed unduly after let-down commences, evacuation of the udder will be incomplete regardless of the length of the milking process or the time spent on machine stripping, i.e. manipulation of the udder and teatcups. Let-down is adversely affected if cows are excited or stressed, and the importance of a regular and quiet routine of washing and preparation cannot be overemphasized. Cows are creatures of habit, and any change or upset will disturb their conditioned reflexes and milk ejection, and hence their milk production. Poor let-down can become habitual with retention of milk in the udder and reduced milk yield. However, some milk, known as the residual milk, is always left in the udder and cannot be removed. This residual milk amounts on average to about 15 per cent of the milk present before milking started, and it does not affect milk yield.

Rate and Frequency of Milking

The pattern of milk flow while a cow is milked is reasonably constant: a brief period of about 1 minute of increasing flow being followed by the period of maximum flow, with a final period of rapidly declining flow. However, the rate at which milk can be removed from the udder varies widely from cow to cow. The difference is due almost entirely to the anatomy of the teats, and in particular to the diameter of the streak canal. Let-down can affect the length of time required to milk a cow but it cannot affect the rate of milk flow, which is determined simply by the bore of the streak canal through which it has to pass. It is thus impossible to train cows to milk either more or less quickly, and slow cows which persistently delay milking should be culled. In general, the higher the yield of a cow, the faster will be the rate of milking, although the overall milking time will be longer. Speed of milking is a highly heritable characteristic, and must be borne in mind when selecting stock for breeding.

The main factor contributing to fast milking—a wide streak canal—can, however, also make the teat more susceptible to the entry of mastitis organisms.

The rate of secretion of milk remains constant for the first 12 hours after milking and declines slowly thereafter. Milking at exactly equal intervals of 12 hours is therefore ideal, but is rarely practised in Britain because of the unsocial hours which it imposes on the milker. The average time intervals are 14·5 and 9·5 hours (i.e. a ratio of 14·5 : 9·5). Experimental evidence suggests that the advantage in milk yield from a 12 : 12 ratio over either a 15 : 9 or a 16 : 8 ratio does not exceed about 4 per cent. The difference between ratios of 14 : 10 and 15 : 9 has not been investigated, but can be assumed to be less than 4 per cent. It seems probable that the importance attached to a 12 : 12 ratio has been exaggerated.

Cows are normally milked twice a day. Experimental results have shown that milking only once a day throughout a lactation compared with twice a day reduces milk yield by 40 to 50 per cent, and omitting one milking per week reduces lactation yield by 5 to 10 per cent. Conversely, milking three times a day instead of twice increases yield by between 5 and 20 per cent.

Milking Machines

When let-down has occurred, the milk is removed from the udder by a machine which applies a vacuum to the end of the teat. To avoid damage to the teat and in order to maintain blood circulation, this application of vacuum is interrupted by rest periods during which the teat is partially protected from the vacuum. This process of interruption is known as pulsation. The vacuum used for milk removal is also used to transfer the milk either to a sealed bucket or jar, or by pipeline to the dairy.

The *vacuum* is created by a pump (which could more accurately be called a compressor) which extracts air from the system and discharges it into the atmosphere. It is important for the vacuum pump to be powerful enough to provide the vacuum required for milking and to generate a sufficient reserve of vacuum to compensate for the admission of air through the application and removal of clusters and milk transfer.

To allow this reserve vacuum capacity to be available when required, a device called a *regulator* is incorporated into the system. This is set to maintain the vacuum at the chosen level, and is normally partially open during milking, opening further to admit air into the system when the vacuum level rises above the desired level and closing when the vacuum level falls too low. Regulators may be operated by spring, diaphragm or weight, but the last is most usual.

To permit a check to be kept on both the level of vacuum and on any undue fluctuations in it, a *vacuum gauge* is incorporated in the vacuum line, preferably where it can be seen by the operator during milking.

Unwanted liquids, such as the circulation wash or milk, sometimes enter the vacuum line, and to prevent them from reaching and damaging the vacuum pump, an *interceptor* or *vacuum balance tank* is fitted into the line near the pump. The interceptor usually takes the form of a removable bucket of about 15 to 20 l capacity, and incorporates a float valve.

To connect the vacuum pump to the milking points there is an *air pipeline* of 25 to 75 mm bore, usually of galvanized steel pipe except where it also forms part of the milking or circulation cleaning circuits, when it will be either of glass or stainless steel. This pipeline is fitted with drain valves at its low points. In cowsheds using bucket units, vacuum taps are fitted to receive the air lines of the units.

The interruption of the vacuum at the teat, termed the *pulsation*, is achieved by means of a valve mechanism termed a *pulsator*, which

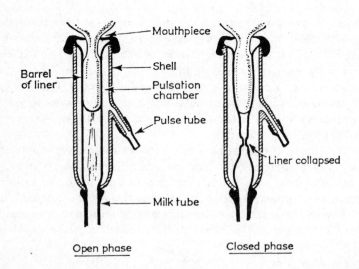

Fig. 10.4. Cross-section of a teatcup, showing the main components and the two phases of pulsation

connects the pulsation chamber (Fig. 10.4) between the teatcup liner and the shell alternately to the vacuum, via a separate pulse tube, and to the atmosphere. When the vacuum is equal inside the liner and in the pulsation chamber, the liner opens, exposing the end of the teat to the vacuum, and when the vacuum drops in the pulsation chamber as air is admitted, the liner collapses under the suction of the vacuum inside the liner. However, the collapse is only partial, and the end of the teat is not completely sealed off from the vacuum, but milk flow ceases and the massaging effect of the collapsing liner keeps the blood circulating and prevents discomfort. Pulsators may be self-contained, each with their own control mechanism (as in most bucket units) or there may be one master pulsator which operates through a number of relay pulsators which transmit the pulse. Pulsators may be operated electronically or pneumatically, and may close all four liners of a cluster at the same time (simultaneous pulsation) or close them in pairs (alternate pulsation).

The Cluster Assembly

This is the complex of equipment which is required at each milking point (Fig. 10.5) and consists of the teatcup cluster with connections for the vacuum and pulse lines and a milk pipe to convey the milk into a bucket, jar or pipeline.

The *cluster* comprises a clawpiece and four teatcups, each made up of a shell, a rubber liner and connecting tubing. The *shells* are usually made of stainless steel and are cylindrical, with the ends adapted to suit the shape of liner. There is a small tube in one side which connects the pulse tube to the pulsation chamber (Fig. 10.4).

The *liner* is usually made of rubber which is either wholly or partly synthetic, since natural rubber absorbs fat and quickly loses its shape and resilience. Each liner comprises a mouthpiece, a barrel and a short milk tube. The mouthpiece should fit the teat tightly enough not to admit air but not so tightly as to cause discomfort. The mouthpiece may be premoulded into shape or formed from a straight piece of tube by inserting a large ring. The short milk tube, which connects the liner to the clawpiece, may be integral with the barrel or separate from it, the joint being made and sealed at the lower end of the shell. It is usually possible to adjust the tension on the liner.

The *clawpiece*, which is usually of stainless steel, connects the four short milk tubes and the four short pulse tubes to the long milk and pulse tubes. The clawpiece incorporates a small milk chamber,

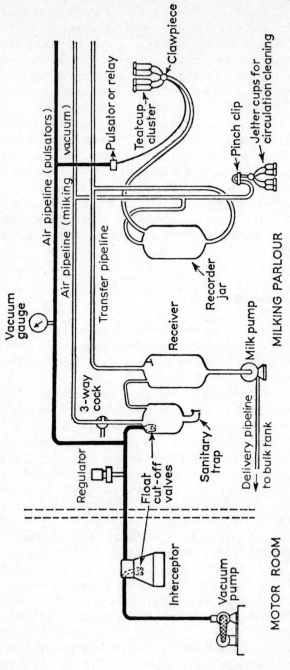

Fig. 10.5. Pipeline milking installation with recorder jar (Based on BS 5545, *Milking Machine Installation*, 1980)

which must be accessible for cleaning, and an air-admission hole of about 0·8 mm diameter which assists the removal of milk from the cluster. The clawpiece must have a minimum capacity of 80 ml below the milk inlet.

Conveying the Milk

In most parlours, the long milk tube from the clawpiece leads to a graduated glass jar of about 27 l capacity, in which the milk is measured, inspected and if necessary rejected, and from which samples can conveniently be taken. While the cow is being milked the jar is under vacuum; to convey the milk to the dairy a valve is turned which cuts off the vacuum, allows air to enter the jar, and opens the connection to the transfer pipeline, which is constantly under vacuum.

In some cowsheds, and in parlours without jars, the milk goes direct from the long milk tube into the milking pipeline, which also carries the vacuum to the teat. In this type of installation, milk yield is measured by meters, either installed permanently or fitted for periodical recording only. In the absence of a permanent meter, there is no way of telling when milk flow has slackened or ceased unless a milk flow indicator is fitted or cluster removal is automated. Blood in the milk is also less easily detected.

In most cowsheds, milking is direct into a bucket of about 18 l capacity, which is carried to the dairy when full and tipped into the milk container.

The transfer pipeline normally terminates in either a glass jar or a stainless steel balance tank. Since this receiver, like the rest of the system, is under vacuum, the milk cannot flow from it by gravity and has to be pumped out against the vacuum by a *milk pump*. This pump is set in operation automatically by either the weight or the level of the milk in the jar. The pump discharges the milk through a pipeline to the bulk container, before entering which it is passed through a *filter* to remove sediment and extraneous matter. A widely used filter is the in-line type, containing a disposable sock of brushed cotton or some similar material through which the milk passes. More usually, however, the filter consists of a pair of non-disposable stainless steel screens. Neither type is wholly efficient in removing all dirt from the milk, and a prime objective of the milking process should be to prevent dirt from entering the milk in the first place. There is a clear seasonal pattern in the incidence of sediment in milk, and it is closely related to the degree of exposure of the cows' udders to mud.

Every machine milking installation should have a standby source of power. This may be by hydraulic drive or direct drive from the power take-off of a tractor to the vacuum pump. This system provides no electricity, and milk pumps and other electrical equipment have to be operated by batteries. The ideal, though expensive, standby is a generator, driven by a diesel engine or tractor, capable of producing enough electricity to run all the equipment. A useful compromise is a device which uses the tractor's power take-off to operate both the vacuum pump and a small generator producing enough electricity for lighting, the milk pump and other small motors.

Producing Clean Milk

The main source of contamination of milk, apart from dirty teats and mastitis, is the build-up of milk residues in the milking installation. To control this build-up requires the conscientious application of a thorough daily cleaning routine.

With bucket machines there is little alternative to manual cleaning, but in milking parlours and cowsheds with pipelines, cleaning is done in place, the equipment being dismantled only when periodic replacement of parts is required. There are two principal methods of in-place cleaning: circulation cleaning, which involves three rinses, and a single-stage cleaning process using acidified boiling water.

Circulation cleaning relies mainly on the properties of the chemicals for its disinfectant effect, but the temperature of the water is also important. The process involves a cold-water rinse, circulation of a hot detergent and disinfectant mixture, and a final cold rinse. The quantity of cleaning solution required is about 14 to 18 l per unit, and the initial temperature must be at least 85 °C. The first 10 l should be run to waste, and the solution then recirculated through the installation, with the pulsators working, for 5 to 10 minutes and finally discharged to waste. The final cold-water rinse may contain sodium hypochlorite for additional protection. In some systems the full routine is done in the morning only, using only a cold rinse with hypochlorite in the evening. To prevent build-up of scale on glass jars and pipes, it may be necessary to draw a proprietary milkstone remover through the installation about once a month.

In the *acidified-boiling-water* (ABW) process, there is no cold rinse, but 14 to 18 l of boiling water are drawn straight from the water heater into the pipeline and discharged to waste. The process takes only 5 to 6 minutes and for the first 2 to 3 minutes 1 l of dilute nitric or sulphamic acid is mixed with the water, in order to reduce the deposition of hard-water salts on the equipment. The acid has no disinfectant effect; this is performed solely by the high temperature of 77 °C reached in all parts of the installation.

If a pipeline is to be cleaned satisfactorily by either system it must be well designed, with no dead ends. For the ABW process, good design and a compact layout are particularly necessary and all components of the installation must be suitable for it. Hypochlorite should be drawn through the installation once a month to prevent a build-up of milk protein.

The ABW process needs water at a much higher temperature, and thus uses more electricity, whereas circulation cleaning requires expensive chemicals. There is no clear-cut economic advantage to either system and both appear to work equally well, though the ABW process is not suitable for installations with long pipe runs. The ABW process is slightly less laborious, taking only 5 to 6 minutes compared with a minimum of 15 minutes for circulation cleaning. In addition to either of these cleaning processes, all rubber parts should be regularly inspected and replaced when they show any signs of deterioration.

Cooling and Storing Milk

Milk is normally collected once a day, by bulk vacuum-tanker lorries from *refrigerated bulk tanks* in the farm dairy. The evening's milk has to be kept overnight, and to ensure good keeping quality it is important that it is cooled rapidly. The Milk Marketing Boards require that milk is cooled to not above 5·0 °C by 30 minutes after the end of morning milking, and the evening's milking is also brought down to and maintained at this temperature. This rapid cooling is extremely effective in controlling the multiplication of bacteria.

Bulk tanks consist of a stainless steel container surrounded by an elaborately insulated jacket, the space between the two being maintained at the correct temperature by chilled water from an ice bank. The refrigerating equipment is usually situated in a separate room. The tank is fitted with a thermometer, from which the cowman and the lorry driver can check the temperature of the milk, and an agitating paddle which ensures even cooling and thorough mixing before milk samples are taken. Tanks are emptied by suction; the lorry driver is responsible for measuring and recording the amount of milk and for rinsing the inside of the tank with cold water. A tap and hose should be available nearby.

For small and inaccessible farms, to which it is impracticable or uneconomic to construct a road capable of carrying a tanker lorry, small mobile tanks are available through the Milk Marketing Boards. These tanks have a capacity of 270 or 500 l and can be towed to a main road by tractor.

In large herds with a production of over about 4,000 l per day, consideration may be given to use of a plate cooler in which the milk is cooled before storage in an insulated, as opposed to a refrigerated tank. A plate cooler is a form of heat exchanger, and will usually use mains water for initial cooling, followed by chilled water cooling to 5·0 °C.

Small bulk tanks can be cleaned by hand, using a long-handled brush, but larger tanks are difficult or impossible to clean satisfactorily in this way and automatic washers are required. These units consist of a water-storage tank, a small electric pump and a metering device for adding chemicals. Instead of rinsing the tank with a hose, the lorry driver presses the starting button of the unit, which pumps clean water through sprinklers inside the lid of the tank, followed by a spray containing iodophor at mains-water temperature. The cowman initiates the rinsing spray before milking. The interior of the tank should be inspected regularly, and automatic washing supplemented with manual cleaning where necessary.

Maintenance and Testing

Manufacturers supply instructions on the maintenance of their milking machines, but it is not enough simply to hand these to the cowman or leave them in the dairy. Instructions on the weekly and monthly maintenance of the machine should be clearly written down and displayed permanently in the dairy or office.

The following checks must be made before every milking. The vacuum gauge should indicate that the correct vacuum level is reached quickly, and the reading when only one unit is on should be compared with the reading when all are on. If the difference exceeds 2 kPa, the plant should be checked. Other routine checks are to listen for the hiss of the regulator and the sound of the pulsators, also for any air leaks; a constant watch should be kept on all rubber parts and on the air-admission holes in the clawpieces. The interceptor and sanitary trap should be checked and if necessary emptied and washed out.

Every week the oil reservoir of the vacuum pump must be filled with special oil and the belt tension checked to ensure a maximum of 12 mm play at the central point. The regulator and the air inlets to the pulsators also need weekly attention.

At least once a year the whole milking installation should be thoroughly tested by a qualified person. The Milk Marketing Boards offer a service involving an annual visit from a mechanic, who thoroughly checks the equipment and leaves a detailed report for the farmer, who can if he wishes authorize the renewal of minor parts. The test involves a careful inspection and the use of an air-flow meter to check the vacuum capacity of the pump, the air consumption of the various components and the final vacuum reserve. This test shows the efficiency of the pump and the degree of leakage which is occurring; the vacuum gauge and the regulator are also checked. A vacuum recorder is used to test the pulsator and relays, producing graphs showing the pressure changes in the pulse chambers of the teatcups.

The faults most commonly found are insufficient reserve vacuum, faulty controllers and pulsators, and incorrect air-admission holes in the clawpieces.

The Machine and Milking Efficiency

The efficiency of the milk removal process is affected greatly by a number of mechanical factors.

Since machine milking is essentially a process of suction, it is not unexpected that increasing the *vacuum level* increases the milk flow rate and reduces milking time. Unfortunately, as vacuum rises above 50 kPa the yield of strippings also rises, and overall milking time is not reduced; further, if machine stripping is not practised, part of the milk yield will be lost. A reasonable balance between flow rate and strip yield is struck at about 44 to 50 kPa. It is essential to have sufficient reserve vacuum capacity to cover the admission of air, to help maintain a stable vacuum and to avoid any tendency for the clusters to fall off.

Pulsation can be varied in two ways. The *pulsator ratio* can be changed, i.e. the ratio of the times during which the liner is open and closed. If the liner is open for twice as long as it is closed, the ratio is

67 per cent. Milk flow rate increases as the pulsator ratio widens, within certain limits; ratios between 50 and 67 per cent are recommended. The *pulsation rate*, i.e. the number of cycles per minute, can also be varied, but extremes are to be avoided, the recommended rate being from 45 to 60 cycles per minute.

Liner design also influences milking efficiency. The ideal liner should remove the milk quickly, with a low strip yield and with the minimum tendency to admit air or to fall off the teat. The liner should also be comfortable and not tend to crawl up the teat during milking. It is thought that the design of the liner barrel affects milking rate, while that of the mouthpiece governs strip yield. Generally, stretched liners are better than moulded liners, and hard mouthpieces and wide-bore barrels should be avoided.

Strip yield, and the need for machine stripping, can be reduced by the *weight of the cluster*, but as the weight increases there is more tendency for the cups to slip off the teats. Recommended cluster weights are from 1·5 to 3·5 kg. If the main weight of the cluster is in the clawpiece, the distribution of weight between the four teats will depend on udder shape and on the 'hang' of the milk tubes; a better weight distribution is achieved by having the main weight of the cluster in the teatcups.

Milking rate is also affected by the *height of the recorder jars* in relation to the udder. There is a slight advantage in having the jars below rather than level with the udder, but a significant disadvantage in having the jars higher than the cows' backs; the latter arrangement can also cause considerable fluctuation of vacuum at the teat. Where milking is direct into a pipeline, the bore, length and design of the pipeline have a major effect on vacuum stability and milking rate.

Milking Machines and Mastitis

Design, maintenance and operation of the milking machine all affect the incidence of udder disease. The recommendations given previously on design and maintenance, and in Chapter Eleven on operation, all relate to this problem. Most udder infections occur at milking time through bacteria penetrating the streak canal of the teat, and the control of mastitis involves efforts to reduce this penetration and also to reduce physical damage to the teats, both external and internal.

Vacuum level and pulsation rate and ratio, if within the limits recommended, do not significantly affect udder infection, but undue fluctuations of vacuum can cause problems. Measures to reduce this fluctuation include an adequate vacuum reserve, correct choice of liner, keeping air-admission holes in the clawpieces clear, positioning of jars and design of pipelines, and efficient handling of the milking units.

The importance of establishing a good and regular milking routine has already been emphasized, and in general efficient milking and good udder health tend to go together. The teatcups should be put on to a clean and well prepared teat and should be removed as soon as possible after the end of milk flow. There is evidence that over-milking damages the teat lining, but none that it actually causes infection; it must, nevertheless, be sound practice to avoid over-milking as far as possible. One of the most potent sources of bacterial cross-infection between teats is reverse flow of milk within the cluster; avoidance of this is largely a matter of good design, examples being the fitting of shields in teatcups and one-way valves within claw-pieces. The problem can be alleviated if the vacuum is always cut off before removal of the cluster from the udder. Mastitis is considered further in Chapter Seventeen.

Further Reading

Archer, P., *Milking three times a day*, Report No. 34, 1983, Milk Marketing Board, Reading

Bramley, A. J., Dodd, F. H. and Griffin, T. K. (eds.), *Mastitis control and herd management*, 1981, NIRD-HRI Technical Bulletin No. 4, National Institute for Research in Dairying, Reading and Hannah Research Institute, Ayr

Frandson, R. D., *Anatomy and physiology of farm animals*, 1975, Lea and Febiger, Philadelphia

Milking machine installations, 1980, Part 1, 'Vocabulary'; Part 2, 'Specification for construction and performance'; Part 3, 'Methods for mechanical testing', BS 5545, British Standards Institution, 101, Pentonville Road, London

The mechanization and automation of cattle production, Occasional Publication No. 2, 1980, The British Society of Animal Production, Milk Marketing Board, Thames Ditton, Surrey

Thiel, C. C. and Dodd, F. H. (eds.), *Machine milking*, 1979, NIRD-HRI Technical Bulletin No. 1, National Institute for Research in Dairying, Shinfield, Reading

CHAPTER ELEVEN

Milking Parlours and Cowsheds

Static Parlours—Rotary Parlours—Cowsheds—Elements of the Work Routine—Parlour Performance—Selecting a Parlour—Parlour and Dairy Buildings—Cow Marshalling

In the previous chapter the physiology and the mechanics of the milking process were discussed. It is now necessary to consider the various ways in which the cow and the milking machine are brought together, and the buildings, equipment and techniques which are employed in doing so. There is a huge variety of milking techniques within the industry, from the hand milker in the cowshed to the fully automated rotary parlour in which one man can milk over 100 cows per hour.

The distribution of the various systems of milking, on a herd basis, is indicated in Table 11.1. There is a trend for the larger herds to use a

Table 11.1   Milking systems in 1981

| Milking location and method | (Percentage of total herds) | | |
	England and Wales	Scotland	Northern Ireland
Cowshed			
Hand	0·5	—	3·5
Bucket	14·8	5·5	35·5
Pipeline	20·3	41·0	21·1
Parlour			
Abreast	28·1	—	1·0
Herringbone	33·8	48·5	34·2
Others	2·5	5·0	4·7
	100·0	100·0	100·0

(From *United Kingdom dairy facts and figures*, 1982, Federation of United Kingdom Milk Marketing Boards, Thames Ditton, Surrey)

parlour system; thus a higher proportion of cows is milked in parlours than is suggested by these figures. Parlour milking is practised in 64·4 per cent of all milking installations in England and Wales, 53·5 per cent in Scotland and 39·9 per cent in Northern Ireland.

Static Parlours

Static parlours are so called to differentiate them from rotary parlours, in which the platform rotates.

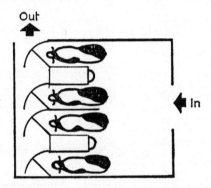

Fig. 11.1. Abreast parlour

The *abreast* (Fig. 11.1) is the simplest form of parlour, and was the first to achieve widespread use—as a movable bail in the 1920s and later in fixed installations. The cows stand side by side in simple stalls, with the recorder jars and equipment mounted as low as possible between each pair of cows. After being milked, each cow can be released individually through the front of each stall, which is formed by a gate. The abreast parlour is cheap and requires only simple building work, but it is impossible to achieve a genuine two-level layout, in which the milker stands in a pit and thus does not have to bend; the best that can be done is to raise the standings by about 400 mm, which is of dubious advantage. The cows entering the parlour have to cross the operator's working area, and owing to the distance apart of the milking units much walking is involved on the part of the milker, though this disadvantage can be alleviated by adopting the back-to-back (Fig. 11.2) or circular layouts.

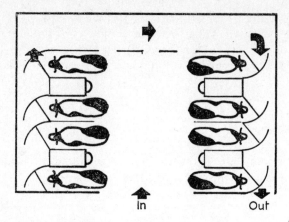

Fig. 11.2. Back-to-back abreast parlour

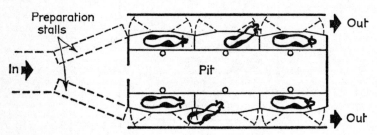

Fig. 11.3. Tandem parlour

In the *tandem* parlour (Fig. 11.3) the milker stands in a pit about
1·5 to 1·8 m wide and 840 mm deep, and the cows stand head to tail
along either side. Each animal has its compartment, which has
individual entry and exit gates with access to a passage running
along the outside of the stalls. In the only version of this parlour now
being installed in this country, there is a high degree of automation:
this includes a preparation stall, mechanically operated entry and
exit gates and automated cluster removal. The automated tandem
allows cows to be treated as individuals, and slow-milking animals
do not hold up other cows as they do in batch-milking parlours.
Tandems have the basic weakness that there is 2·4 m between units
along each side, so that three or at the most four cows on each side is
the practical maximum on account of the distance to be walked by the
milker. Six or eight units are not enough to justify the expensive and

mechanically complex automation. This parlour allows a very short time for eating concentrates.

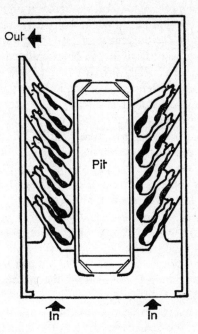

Fig. 11.4.   Herringbone parlour

The *herringbone* (Fig. 11.4) has a similar pit but the cows stand at an angle of about 30 ° to the pit on either side and with their tails towards it. In this way the distance between udders, and hence between milking units, is reduced from 2·4 m to only 900 mm or 1 m and one man can, given automation, handle ten units each side without an undue amount of walking. The jars and pipelines may be mounted above the operator's head (high-level), or beneath the cow standings (low-level), or at an intermediate height (eye-level). The low-level design can only be used for parlours having one stall per unit, but with the other two designs there may be either one or two stalls per unit. High-level jars are associated with slower milking rates and less vacuum stability, and are not recommended for new parlours; most new herringbone parlours are now of the one-stall-per-unit, or 'doubled-up' type. Low-level jars have the advantage over eye-level jars of leaving a completely unobstructed pit, with excellent access to the cows, but eye-level jars are more easily cleaned and drawing off colostrum and reading the graduations on the jars is easier.

In static parlours there may be one milking unit for each pair of stalls (one cow being prepared while the other is being milked) or a unit for each stall. The advantage of providing a unit for each stall lies mainly in increased flexibility in batch-milking parlours such as the herringbone, and a slight increase in available milking time due to the fact that for a brief period all the units may be on the cows at the same time. If the milker is already fully occupied there is no advantage in doubling up a two-stalls-per-unit parlour, but where there is some spare time in his routine, performance may be improved by 10 to 20 per cent. An 8/8 (8 units, 8 stalls, i.e. 4 stalls each side, with a unit for each stall) herringbone, for example, may be expected to give a similar throughput to a 5/10 (5 units, 10 stalls). Most new herringbone parlours have a unit for each stall.

The *multi-sided herringbone* was developed in the United States. The original form was based on a large herringbone, with each side divided into two and the middle of the sides pushed outwards to make a diamond shape. This four-sided version is usually called a *polygon*. The advantages of dividing the number of units into smaller batches are that one slow-milking cow holds up fewer other cows, and that smaller groups of cows move in and out of a parlour better. However, the real advantage is in throughput, an advantage which this type of parlour has over the two-sided herringbone be-

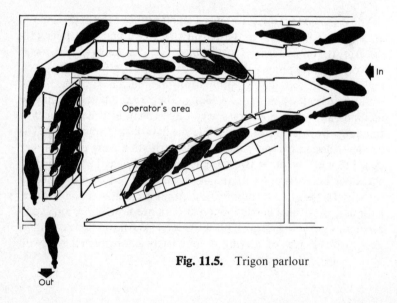

Fig. 11.5.　Trigon parlour

cause of the reduced unit idle time and the increased available milking time.

Although the polygon seems unlikely to find a place in any but very large herds, the three-sided version or *trigon* (Fig. 11.5) may have a wider role to play, though again mainly in large herds. American timings suggest an advantage in throughput from 13 to 20 per cent for the trigon compared with a two-sided herringbone with the same number of stalls.

An idea imported from Holland is the *side-by-side* parlour (Fig. 11.6), in which the cows stand at 90° to the pit, and are kept in place by a bank of self-locking yokes. The jars are housed beneath the cow standing and the teatcups are placed between the back legs of the animals.

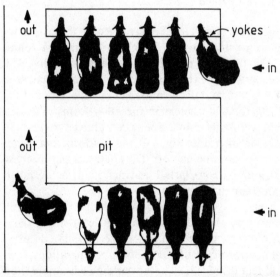

Fig. 11.6. Side-by-side parlour

Rotary Parlours

In rotary parlours, the cows stand on a circular revolving platform and the milker or milkers may be either inside or outside this circle. The cow is normally taken through the whole milking routine in the course of one rotation of the platform. The cows can be arranged on the platform in three ways: head to tail (tandem), at an angle (herringbone) or with their heads to the centre (abreast).

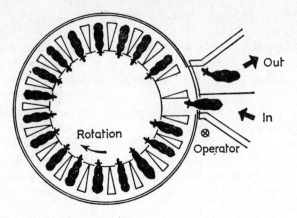

Fig. 11.7. Rotary abreast parlour

Rotary abreast parlours (Fig. 11.7) are arranged so that the cows walk on to the platform but have to back off it and turn round before going out, but this does not appear to cause any difficulty. The operator stands outside the circle, and can thus easily encourage slow cows on to the platform, but the milking units move out of his sight as the platform rotates, whereas in the other forms of rotary parlour they remain within his vision. The rotary abreast may rotate either continuously or intermittently, i.e. the platform may stop briefly to allow cows to move on and off. The abreast is the cheapest form of rotary parlour both to install and maintain, being much simpler mechanically than the other types.

Rotary parlours may have 6 to 40 or more stalls. Operation is usually by one man up to about 12 stalls; above that number there will be one man at the entry point with either a second man or automated equipment performing the end-of-cycle operations, i.e. cluster removal and teat disinfection. Large parlours with more than about 25 stalls will have two men at the entry point, one washing and preparing the udders, and the other putting on the clusters.

The advantages of the rotary over the static parlour are that it saves the time spent in changing batches of cows, and that automation of other elements of the work routine is easier because each cow has to pass fixed points on the way on and off the platform. However, lack of confidence in the efficiency of the automation of the end-of-cycle operations has led to most rotary parlours having a man at the exit as well as at the entry point. This reduces the throughput per man-hour to no more than can be achieved in a static herringbone without automation, and the high extra cost of the rotary cannot in these circumstances be justified. Other problems with rotary parlours are poor cow flow on to the platform, mechanical breakdown, and heavy maintenance costs.

Cowsheds

In cowsheds the cows are housed and milked in the same stalls. Each cow is secured by a neck chain or yoke, which allows access to a manger and water bowl. The cows usually lie in pairs, with a division between each pair and access for the milker between the cows of each pair.

Where milking is by bucket machine, the milk must be carried to the dairy, where it will usually be tipped into a hopper filter mounted in a hole in the lid of the bulk tank. The main differences between milking in a cowshed and in a parlour are first, that in the cowshed the milker operates on the same level as the cows, and second, that the units are brought to the cows instead of *vice versa*. A reasonable minimum performance in a cowshed using bucket machines would be 30 cows milked per man-hour; with a fixed milk pipeline discharging into a bulk tank, the rate should be increased to between 40 and 60 cows per man-hour.

The main reason for the change from cowsheds to parlours lies less in the milking than in the other operations such as feeding, littering, dung removal and tying and untying the cows.

Elements of the Work Routine

The work routine in milking comprises the various operations which the milker has to perform on each cow. These are as follows.

Letting cows in and out: this is done by batches in the herringbone, but individually in the tandem, abreast and rotary. Cows move more freely into parlours when in batches, but movement out of long herringbones can be slow. Cow entry and exit gates in herringbone parlours should be capable of being operated mechanically from anywhere in the pit. A good cow flow depends on a good relationship between cows and milker and on good design of collecting yard, cow entrance and exit.

Feeding: most cows are given concentrates in the parlour, and are usually identified and rationed according to yield, though feeding by groups may be practised in large herds. In abreast parlours, concentrate feeding is often by hand, using a shovel from a bin or hopper; but in herringbone and rotary parlours, mechanical and automated systems are usual.

The simplest feeder is a fixed hopper, with a metering device operated by a lever from the pit. The most common type of feeder in herringbone parlours involves identifying each cow on entry, referring to a chart showing the amount of feed she is to receive, and setting this amount on a dial. When a dial has been set for each cow in the batch, a button is pressed and all the cows on one side of the parlour are fed simultaneously. In a more complex and expensive type of feeder, each cow's ration is programmed periodically into a small computer and the operator need only tap out the number of the cow on the panel. A still more sophisticated system, which is in only very restricted use, is fully computerized, and identifies each cow from a device hung round her neck; it then both feeds the cow according to yield and advises the milker of any attention which she needs.

Udder washing: efforts to automate this operation in rotary and tandem parlours have had only limited success. For reasons of hygiene and convenience, the bucket and udder cloth should be replaced in all parlours by a warm-water udder-wash spray. This supplies water, into which disinfectant may be metered, at the correct temperature through several hoses which hang down into the pit at convenient points. Washing is sometimes omitted altogether, but there is a clear obligation on the milker to ensure that the teats are clean when the cups are applied. After washing, the teats should be dried with disposable paper towels; tests have shown that washing the teats without drying them shows no reduction in bacterial contamination of the milk compared with not washing them at all. The best way of reducing the time spent on udder washing is to ensure that the cows lie clean.

Taking the foremilk: this involves squeezing a few drops of milk from each teat into a special cup in order to ensure that the teat orifice is not blocked, to remove the first milk, which may have a high content of bacteria, and to allow the milker to check the milk for mastitis, blood and other abnormalities.

Attaching the cluster: this is the element of the routine which seems least likely to be successfully automated.

Removing the cluster: as soon as milk flow has ceased, the vacuum should be cut off from the cluster and the teatcups gently drawn off the udder. This process has now been successfully automated, the cluster usually being suspended by a cord which, after the vacuum has been cut off and after a pause, draws the cluster off the teats upwards and towards the pit. The process may be initiated by the milker when he sees that milk flow has ceased, or it may be fully automated, the end of milk flow being detected by the sensor of the cluster removal device. Automatic cluster removal (ACR) devices are now in widespread use in all types of parlour. They slightly reduce the work routine time and take all decisions on cluster removal from the cowman, leaving him free to operate a smooth routine and to handle more units. ACR also avoids the danger of overmilking.

Milk transfer: a jar partially filled with milk has to be emptied before the next cow is milked into the same jar. This can cause delays, particularly in parlours with two stalls per unit, and milk transfer is now frequently linked to either automated or semi-automated cluster removal and performed automatically. In rotary parlours, milk transfer can be initiated by the rotation of the platform as it passes the cow exit point.

Teat disinfection: an essential element in the mastitis control routine is the immersion of each teat immediately after milking in a cup containing a suitable iodophor or dilute hypochlorite. This operation, termed teat dipping, should be included in every milking routine, and it is helpful to have several cups of teat dip conveniently placed round the pit. Hand-operated sprays may be used as an alternative. Efforts to automate teat disinfection, using spray nozzles mounted in the floor and sides of an exit race, are showing promise.

Parlour Performance

The milking performance of a parlour can be measured in either litres of milk per man-hour or cows per man-hour, but the latter measure is the more useful since daily milk yield fluctuates according to the stage of lactation, whereas the time which the milker spends on each cow remains virtually constant throughout the year. However, a high throughput of cows should never be pursued at the expense of good stockmanship.

The work-routine time (WRT) is the total time taken to perform for each cow the various elements previously described, and hence the maximum number of cows which can be milked per hour is found by dividing the WRT per cow into 60 minutes. Three typical work routines, with timings, are given in Table 11.2. The total WRT in the examples varies from 1·0 minute for a full routine to 0·5 minute in a rotary parlour with ACR, giving maximum throughputs of 60 and 120 cows per hour respectively. The WRT can only be reduced by either automating or omitting certain elements of the routine.

Table 11.2. Work-routine times (WRT) in parlours

| | | (minutes) | |
| | | Herringbone | Rotary With |
Elements in work routine	Herringbone	with ACR	ACR
Change batch and feed	0·20	0·20	auto
Take foremilk, wash and dry	0·25	0·25	0·25
Attach cluster	0·20	0·20	0·20
Remove cluster	0·20	auto	auto
Dip teats	0·08	0·08	auto
Miscellaneous	0·07	0·07	0·05
Total	1·00	0·80	0·50
Cows milked per hour	60	75	120

The potential maximum number of cows cannot be milked per hour unless the parlour is designed so as to give each cow sufficient time in which to milk out. In general, adding more units to a parlour increases the time available for milking out. Each unit added to one side of a parlour will increase the total WRT for that side by one WRT, and the objective in planning a parlour is to have sufficient units so that the milker just has time to perform his routine on one batch of cows while the other batch is milking out. It is thus important to know the time taken for the average cow in a herd to milk out, and this time is related to milk yield (Fig. 11.8).

It is usual to calculate the average required milking time from the mean peak yield of the herd at morning milking. If the average herd yield is 5,000 kg, half of this yield is given in the first 100 days, with an average yield of 25 kg per cow per day. Using a typical ratio of 1:1·5 for the yield at evening and morning milkings, mean peak yield at morning milking will be 15 kg. From the equation used to produce Fig. 11.8 a milking time of 5·9 minutes is required for this

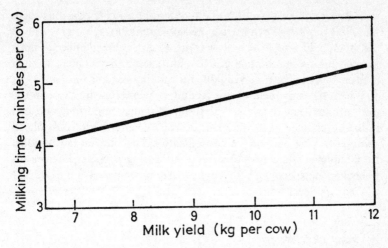

Fig. 11.8. Relationship between milking time and milk yield for 1,934 cows in 28 herds
(From *Machine Milking*)

yield. The data in Fig. 11.8 should only be used as a guide with existing milking installations, but it would appear that the relationship exists up to milk yields of 15 kg per cow. The equation on which Fig. 11.8 is based is: $t = 2 \cdot 75 + 0 \cdot 207 \, y$, where t = milking time in minutes and y = yield of milk in kg.

Table 11.3. Mean milk yield (kg per cow per milking) up to which maximum performance is possible in various types and sizes of parlour

Work routine (*minutes per cow*)	Cows per man-hour	*Static herringbones*									
		4/8	*8/8*	*5/10*	*10/10*	*6/12*	*12/12*	*7/14*	*14/14*	*8/16*	*16/16*
1·2	50	11	14	14	16	16	18	18	20	20	22
1·0	60	9	11	11	14	14	16	16	18	18	20
0·8	75	7	9	9	11	11	14	14	16	16	18

		Rotary herringbone and abreasts			
		12/12	*14/14*	*16/16*	*18/18*
0·6	100	11	14	16	18
0·5	120	9	11	14	16

(Based on a table prepared by A. J. Quick, ADAS)

The level of milk yield at one milking up to which various types and sizes of parlour can realize their potential throughput is given in Table 11.3. This table also shows that increasing the number of units beyond the optimum number does not increase throughput, though it does increase the time available for milking out and eating.

Where the only facilities for feeding concentrates are those in the parlour, the time available for eating becomes important and can influence choice of parlour. However, it is better where possible to arrange to give part of the concentrate ration outside the parlour, thus reducing the eating time required during milking. Factors influencing the speed with which cows eat concentrates are discussed in Chapter Seventeen.

Selecting a Parlour

The first decisions to be taken are the number of men to be engaged in milking and the length of the milking time. Most parlours being installed in Britain at present are of the herringbone type, and most of them are designed for operation by one man. There is no definite limit on the length of a milking period, but it does not normally exceed about 2 hours with a maximum of perhaps 2·5 hours at peak periods. It is, however, emphasized that there is nothing rigid about these times, which are often extended where shift milking systems are employed.

It is not unusual to select a parlour in which the full number of cows in the herd can be milked in about 2·25 hours. Conversely, a herd is often expanded to 2·25 times the potential hourly throughput of the parlour, e.g. 135 cows for a 10/10 herringbone parlour with a throughput of 60 cows per hour. This calculation makes allowance for the fact that for most of the year part of the herd will be dry.

If more than 60 cows are to be milked per hour, the work routine time will have to be reduced below 1·0 minute. In a static herringbone, cluster removal and milk transfer may be automated, thus reducing the WRT (Table 11.2) to 0·8 minute and giving a potential throughput of 75 cows per hour. It will be seen from Table 11.3 that in order to achieve this throughput at, for example, mean peak yields at morning milking of 14 to 16 kg, the selected parlour must have either seven milking units and 14 stalls (7/14), or eight units and 16 stalls (8/16). If the units are 'doubled up', there will be either 14 units and 14 stalls (14/14) or 16 units and 16 stalls (16/16). It is difficult to foresee scope for reducing the WRT in static parlours below about 0·7 minute, which would give a maximum throughput of 85 cows per man-hour, although higher throughputs have been recorded in polygons and trigons.

In a rotary parlour, the WRT (Table 11.2) can be reduced to 0·5 minute, making possible a throughput of 120 cows per man-hour, but the operator can only milk 120 cows if the platform delivers 120 cows to him per hour. The number of cows delivered by the platform will be the number of stalls multiplied by the number of rotations of the platform per hour. Platform rotation time cannot be reduced without also reducing the time available for the cows to milk out and eat, so that it is essential to have sufficient stalls. The impracticability of extending a rotary parlour once it has been installed reinforces this point. Table 11.3 shows that with a WRT of 0·5 minute a rotary parlour should not have less than 16 stalls if the typical mean peak yields of 14 to 16 kg are to be obtained.

Although the throughput of a parlour will depend on the individual operator and the work routine employed, a rough guide to typical performance in selected types of parlour in average conditions is shown in Table 11.4.

Table 11.4. Typical performance in selected types of parlour

Type of parlour	Number of milking units	Number of stalls	WRT (minutes)	Cows milked per man-hour
Abreast	4	8	2·0	30
Herringbone	4	8	1·5	40
Herringbone	10	10	1·0	60
Herringbone with ACR	16	16	0·8	75

Parlour and Dairy Buildings

Milking parlours are the most used buildings on a dairy farm, and must be substantially built, with a good finish. A good, though expensive material for the internal walls is a concrete block which has a hard, glazed surface integrally bonded to one side. The alternative is to render and paint the walls; chlorinated rubber paint is recommended. Corners of walls round which cows have to turn should be rounded to prevent injury, and floors must have a non-slip finish. Parlours should be well ventilated.

Concentrates may be stored either in a loft over the parlour, into which they are delivered from a bulk lorry, or in a free-standing bulk bin, from which they are taken to the feed dispensers in the parlour by augers or conveyors. The systems do not differ greatly in cost. The loft floor is usually flat; the underside of the joists supporting it should not be lined with a ceiling, which can provide an ideal harbourage for vermin, but the floor must be completely dustproof.

Dairies should be large, since herd numbers tend to increase and milk yields to rise, and there may be need to accommodate a second or larger bulk milk tank. Adequate space is required around bulk tanks to permit easy cleaning; 600 mm is a normal minimum. At least 1·2 m should be allowed at the end of the tank where the milk pump is situated.

Parlour and dairy buildings should always include an office of adequate size in which the herdsman can keep his essential records and veterinary and other supplies. If the dairy unit is some way from the other farm buildings, it should include a WC and washbasin and facilities for making a hot drink.

Good lighting is essential in both parlours and dairies. Fluorescent lights are ideal above parlour pits, and with low-level jars it is worth considering installing bulkhead lights between the jars. In dairies there must be good lights above each end of the bulk tank so that the whole of the interior is well illuminated. If there is any danger of frost affecting the milking equipment, heaters should be provided in the parlour; these can be operated by time switch or by thermostat. Wiring in parlours and dairies should be in plastic conduit, properly earthed and tested, and never left unprotected in situations where it can be attacked by vermin or otherwise damaged.

Cow Marshalling

The most lavishly equipped parlour will not fulfil its potential unless cow flow in and out is smooth and efficient. It is now believed that cows move better in a collecting yard if they are not packed too tightly; a recommended allowance for a Friesian cow is 1·2 m². Circular collecting yards, although popular in New Zealand, have made little impact in this country. They consist of a circular yard in which a gate is hung on a central post, on which it rotates (propelled by a weight or a motor), thus steadily reducing the area available to the cows. These yards are not easy to fit into the more compact and often fully covered layouts found on British farms. They are difficult to clean with a tractor scraper.

Rectangular collecting yards should be long and narrow rather than square, and the cows should enter them at the end farthest from the parlour. The yards should slope away from the parlour, and should preferably be at least partly covered. Whether the cost of covering the collecting yard can be justified will depend on how the cows are housed, but the yard should always be well sheltered from wind.

To comply with the regulations, there must be a door between the parlour and collecting yard unless the latter is to be kept to the same standard of cleanliness as the parlour. Where even only part of the collecting yard is covered, there should be no wall between the yard and the parlour, but either an up-and-over garage door or side-hung sheeted gates which can be closed between milkings. Cows move into a parlour far better if they can see the whole of the interior; entry into the parlour should be straight and not at right angles.

Cows in the collecting yard can be brought towards the parlour by an 'electric dog' consisting of an electrified wire or pipe which is winched towards the parlour as milking proceeds. This device frightens the timid cows at the rear of a group and has no effect on the slow cows at the front, and is not recommended. A better, but much more expensive, alternative is a tubular steel gate suspended from rails along each side of the yard and driven by a motorized wheel. A cheap and effective method is to mount a pair of gates about one-third of the way down the collecting yard from the parlour. These lie flat against the walls until near the end of milking, when they are latched together, thus reducing the amount of room available for the remaining cows.

Exit from herringbone and tandem parlours may be either through the end or a side wall. There is some advantage in the latter arrangement, since the cows will, in any layout, have to turn through a right angle at some point. If this turn is inside the parlour the cows will all pass through the same exit and the milker can conveniently divert any cow required for further attention. Even in the simplest buildings, there should be facilities for diverting cows into a holding pen, which should be immediately outside the parlour exit door and on the cows' normal return route. Diversion of individual cows is easy in a rotary parlour, since the cows come off the platform separately, but in herringbone parlours it is better to accept that the milker must go to the exit and select the particular cow.

Flies can be a source of irritation in summer to both cows and milkers, and it is worth making efforts to reduce fly numbers by eliminating suitable breeding grounds near the parlour and by discouraging entry of flies into the milking premises. Various methods can be employed at the parlour entrance, including a screen of plain water dispensed by spray booms or perforated hose, or plastic strip curtains. Fogging with insecticide from aerosol generators can also be effective, but must not be practised indiscriminately.

Further Reading

Choice of milking parlour, Booklet 2426, 1982, Ministry of Agriculture, Fisheries and Food, London

Herringbone, trigon and polygon parlour milking, Booklet 2411, 1982, Ministry of Agriculture, Fisheries and Food, London

The mechanization and automation of cattle production, Occasional Publication No. 2, 1980, The British Society of Animal Production, Milk Marketing Board, Thames Ditton, Surrey

Thiel, C. C. and Dodd, F. H. (eds.), *Machine milking*, 1979, NIRD-HRI Technical Bulletin No. 1, National Institute for Research in Dairying, Shinfield, Reading

CHAPTER TWELVE

Buildings

Housing Since 1945—Strawyards—Cubicle Houses—Construction of Cubicle Houses—Environment—Cubicle Design—Cubicle Beds—Feeding Areas—Silos—Cow Handling—Calving and Isolation Boxes—Young Stock—Bull Pens—General Layout—Planning Regulations and Appearance

The cow is a tolerant creature and can accustom herself to a wide range of types of housing with little apparent effect on her performance. This does not imply that buildings and their design are not important, but emphasizes that milk can be produced efficiently in cheap and simple buildings if the basic principles of housing are followed.

In fact, most cow buildings are planned with the comfort and convenience of the man in view rather than the cow. This is acceptable, provided that a few simple points concerning the cow's requirements are observed. Firstly, cows do not mind being cold, and they do not particularly mind being wet, but if they are both cold and wet, and are subjected to wind or draught as well, they will suffer. Secondly, cows dislike being cramped for space.

Housing Since 1945

Forty years ago, nearly all the cows in Britain were kept in cowsheds, also called shippons or byres, in which they were tied by the neck, fed and milked. Between 1950 and 1960 the yard-and-parlour system was introduced, in which the cows were no longer individually secured, but were loose-housed in a strawyard and milked in a special milking parlour which held only a few cows at a time. The two great advantages of the new system over the cowshed system were the removal of the rigid limitation of numbers which was imposed by the cowshed, and the reduction in labour in both milking and bedding the cows.

The major disadvantage was that cows could no longer be rationed accurately as individuals except for the amount of concentrates given in the parlour. The cowshed is still popular in certain areas, with a few completely new cowsheds being built, but in general cowsheds have been outdated by more modern and less laborious systems. The loss of individual rationing is a real disadvantage, but with the trend to larger herds this was likely to prove impracticable, and modern dairy farming is largely concerned with obtaining maximum milk production without individual rationing of forage.

Since 1960, the strawyard has been largely superseded by cubicles, which are small, individual raised beds into which the cow can walk forwards but not backwards and in which she has just enough room to lie comfortably but not enough in which to turn round. The cow is not secured, and can go in and out of the cubicle at will. If the dimensions are correct, dung and urine fall in the concreted passage and the bed remains clean. As a result, cubicles require less litter and keep cows cleaner than do strawyards; this is particularly true in late winter, when mastitis can be difficult to control in strawyards. In cubicle houses, cows on heat do not disturb other animals, and there is less danger of trampled teats. About 80 per cent of all new dairy housing is in cubicles.

Strawyards

Friesian cows in strawyards require an area of about 5·5 to 6 m² if they are to keep reasonably clean. This assumes that the cows will stand on concrete to feed, and that this concrete will be scraped daily. Since each cow needs 600 mm of manger space, the width of the building each side of the manger needs to be about 9 to 10 m, of which 3 m will be concreted and the remainder strawed. To allow for the build-up of dung over the winter it is desirable, if soil conditions allow, to lower the floor of the bedded area about 600 mm below the feeding passage, with steps up for the cows (Fig. 12.1). The bedded area need not be concreted, but a layer of rammed chalk will make cleaning-out easier. It is useful if the cows have access to the feeding passage along its full length, to reduce trampling in one area, which greatly increases straw use. To facilitate feeding while the cows are in the yard, it is advisable to have a central drive-through passage with strawyards.

Buildings for strawyards should have a minimum headroom of

about 3·4 m for ease of cleaning out, and the walls should be strongly built to resist the thrust of both the accumulated dung and of the foreloader during cleaning-out. If concrete blocks are used, they should be 225 mm thick, filled with concrete and with some rein- forcement, and tied to the stanchions of the building. If the bays are 4·5 m or more, there should be an intermediate pier in each bay. When building a new strawyard, it is wise to allow for possible future conversion to cubicles. This will affect the width of the building and the position of gateways, and ventilated ridgepieces should be fitted to the roof. The correct place for a water trough in a strawyard is between the feeding passage and the bed, with access to the trough only from the concreted side. In this position the two problems of dung build-up and excessive use of litter round the trough are avoided.

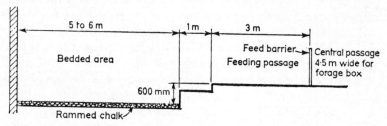

Fig. 12.1. Cross-section through a strawyard

Cubicle Houses

The main structural differences between strawyards and cubicle houses are that the latter can have lower eaves heights and walls of very light construction, since there is no build-up of dung, no use of foreloaders, and the cows do not come into contact with the walls. A further difference is that since the cows must occupy cubicles in all parts of the building, it must be equally comfortable throughout. This means cladding the whole, or at least the greater part of the walls, which in turn makes good ventilation essential.

There are four main types of cubicle house: the kennel, the pre- fabricated house, the dutch-barn type and the clear-span building.

The *kennel* consists of two ranges of simple timber frames, usually clad with corrugated iron on the roof and sides, facing each other across a passage 2·4 m wide and with a gap of about 300 mm between the roofs. If the concrete floor is laid to a longitudinal fall of not less than 1 in 70, the corrugated roof sheets can be laid longitudinally and carry the rainwater to the lower end, thus needing only a short length of gutter. Cows are comfortable in kennels, which provide them with a warm lying area with immediate access to plenty of fresh air. The buildings are cheap, and particularly suitable for erection by farm staff, but are easily damaged by tractors. Kennels are normally associated with a feeding area which is at least partly uncovered, and this can entail much foul run-off (Chapter Thirteen).

The *prefabricated cubicle house* varies widely in type of construction, but in general the roof is supported by carrying up some or all of the front uprights of the cubicle divisions. This house frequently incorporates a central, covered feeding area. A building of this type avoids the foul-run-off problem, and at its best it can provide adequate housing at comparatively low cost, though it is usually difficult to convert to any use but the housing of cows. In some prefabricated buildings the size of both cubicles and passages is too small.

The *dutch-barn* type of building has a central feeding area, with a lean-to each side accommodating cubicles. This is normally the cheapest way of providing a building which is more substantial than most prefabricated houses, and with some measure of adaptability for other purposes.

The *clear-span building* can accommodate virtually any combination of cubicles and feeding arrangements. This tends to be the most expensive type of building, but it retains flexibility of use, being suitable for conversion for either other livestock or storage. It has no immediate functional advantage over the dutch-barn type and its only advantage over the prefabricated type is that it requires less maintenance. Cows are equally comfortable in any of the four types of building provided that space is adequate and the ventilation correct.

Construction of Cubicle Houses

Clear-span and dutch-barn type buildings may be of timber, steel or concrete. Timber must be pressure-treated with a preservative to give immunity from attack by insects and a long life even when exposed to wet conditions. It is easy to cut, join and fix to, and ideal for buildings to be erected by farm staff. It is, however, less economic for spans over about 12 m, when either complicated trusses or laminated portal frames have to be used. Steel, while somewhat less easy to fix to than timber, has a very high resistance to impact. On the other hand it does corrode, though during the expected life of most livestock buildings this is more likely to affect appearance than stability. Wide spans are no problem with steel. Concrete has the advantage that it requires no maintenance, but despite the provision of numerous fixing holes in the stanchions, it remains less convenient for fixing to than steel or timber. For smaller-span buildings concrete is expensive, but for the larger spans it can be competitive with steel and timber.

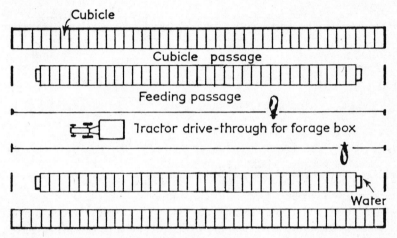

Fig. 12.2.   Basic cubicle-house layout

The layout of cubicle and feeding areas is affected by two main factors: the convenience of being able to scrape the slurry from the one while the cows are in the other; and the desirability of dividing a herd of more than about 90 cows into two or more groups. Both these requirements are admirably met by the basic cubicle-house plan (Fig. 12.2). This design is difficult to improve, since cow flow from either side of the central feeding area to and from the parlour is simple, and the cows are easily held in either their sleeping or feeding areas. Long runs of cubicles without cross-over passages should be avoided. This type of house can be extended indefinitely, since the width of one cubicle (of which there are two opposite to each other) just equals two manger spaces. The main limit to length is the distance which slurry has to be scraped.

An alternative plan (Fig. 12.3) has feeding areas along the outside of the cubicle house, with feed delivered from roadways outside the

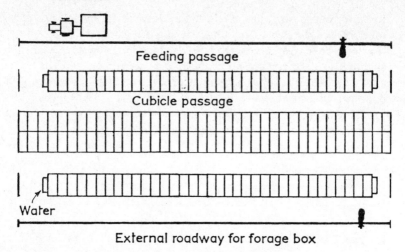

Fig. 12.3. Alternative cubicle-house layout with external feeding

building. This layout is simple and superficially attractive, but can produce a draughty building unless carefully designed.

Although a reasonable fall on a site is an advantage, a steep slope can pose problems. One solution is simply to bulldoze the site level, but this tends to leave steps at the top and bottom of the site, which may make future expansion more difficult and expensive. It is therefore important, before deciding to level a site, to consider how further buildings could be fitted on to the proposed layout. There is no disadvantage in a longitudinal fall of 1 in 40 on a cubicle house and, where there is no alternative, falls even steeper than 1 in 30 are possible, though not recommended. Cross-falls can also be accommodated by placing the rows of cubicles at different levels. Cross-passages between cubicle and feed passages should be raised about 150 mm to prevent slurry from being scraped into them during cleaning of the passages. These cross-passages are the ideal place to site drinking troughs.

Instead of solid-floored passages, it is possible to have passages formed from precast concrete slats, about 150 mm wide and 165 to 190 mm deep, with a space of 40 mm between them. The dung is trodden through the gaps and falls either into a channel or into a cellar below, which must have flat, smooth and waterproof floors and walls.

The cost of slatted passages is high, but they have the advantage that no scraping is needed. Slatted passages are far more popular in Scotland than in England and Wales.

Environment

A cubicle house does not require an artificially controlled environment, i.e. fan ventilation, but must offer the cows sufficient protection from the weather to allow them to lie with comfort in any cubicle. If a gate in one end of the building has to be open, the other end of the building should be well sealed.

Eaves height and roof pitch have a considerable effect on the atmosphere in a cubicle house. Buildings which are too high are cold, and the eaves height should be just high enough for the gateways to admit a tractor and cab, i.e. about 2·8 m. Where a steep longitudinal slope on a long building would give an eaves height at the bottom end of more than 4 m it is worthwhile to step down the roof halfway along its length.

Ventilation is of crucial importance in cubicle houses. The system can be simple, but it must be adequate and matched to the density of stocking and to the situation of the building. There must be an outlet at the ridge, which may be either an open ridge or ventilated ridgepieces. In addition there must be inlets at the sides, which are usually either an eaves gap of about 150 mm or open-spaced boarding. Additional ventilation may be achieved by raising the lower ends of a row of sheets halfway up each slope of the roof, leaving a gap of about 50 mm. Ventilation must be watched particularly carefully when existing buildings with sealed ridges are converted to cubicles.

Cubicle Design

The design of cubicle divisions is currently of two main types: the Newton Rigg (Fig. 12.4) or the simple, two-rail type, which may be of either steel or timber (Fig. 12.5). There are no advantages, and some disadvantages, in having three rails rather than two.

For Friesian cows, the length of the cubicle from the head end to the passage side of the kerb should preferably be 2·20 m and certainly not less than 2·13 m. There is no disadvantage in a longer bed, provided a headrail is used. The width from centre to centre of the divisions should be 1·20 m to 1·25 m, although in older installations narrower dimensions, such as 1·14 m, are frequently found.

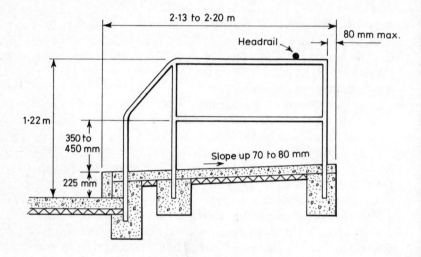

Fig. 12.4. Newton Rigg design for cubicle division

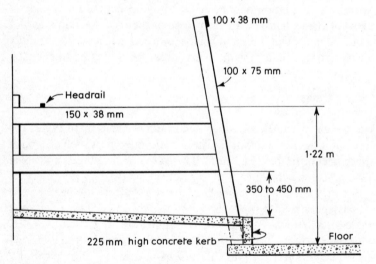

Fig. 12.5. Timber cubicle division

The top of the cubicle division should be about 1 m above the bed, to make it difficult for cows to stand diagonally in the beds, and should extend to within 300 mm of the passage to prevent smaller cows from walking along the kerbs and backing into the beds. The lower rail should be between 350 mm and 450 mm above the bed. If this rail is too low, legs can become trapped, whereas if it is too high, cows may roll underneath and frighten or hurt themselves when rising. It is desirable to slope the rear upright away from the passage.

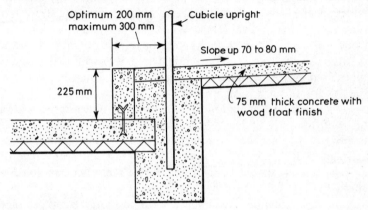

Fig. 12.6. Detail of concrete cubicle kerb

The kerb (Fig. 12.6) should be 225 to 250 mm high to prevent slurry being pushed on to the bed during scraping. It should not project above the bed as a lip except where sand is used as litter, when a lip 50 to 75 mm high is required. This lip should either be wide enough to encourage a cow to stand on it (250 mm) or narrow (75 to 100 mm).

Cubicle Beds

The bed of the cubicle, i.e. the base upon which the litter is laid, must be firm enough not to break up into lumps which are most uncomfortable for the cow. Attempts to achieve this with rammed chalk and similar materials are rarely successful and a concrete bed is to be preferred. A thickness of 50 to 75 mm will suffice if the base is well consolidated, and a wood-float finish is recommended. With this thickness of concrete, insulation of the bed has little effect. The bed should slope up 70 to 80 mm away from the passage. Tarmac has been used successfully instead of concrete and is claimed to have advantages in comfort and warmth. Rubber and plastic mats are also used, but they do not dispense with the need for litter. Whatever the surface, it must be generously and frequently littered to provide a clean, hygienic and comfortable bed for the cows. Attempts to dispense with or significantly reduce the amount of litter have not been successful.

Straw makes good litter; chopped straw reduces the amount pulled off the beds by the cows' feet compared with long straw and suits many slurry systems better. Wood shavings and other forms of wood waste are also useful, although there may be an association between sawdust and *E. coli* mastitis. Diced newsprint is also sometimes used. Soft sand makes excellent, hygienic litter, but it is not suitable for all slurry systems.

Cubicle passages should ideally be 2·30 m between kerbs, with a range of from 2·17 to 2·44 m. Cubicles can be fitted into a narrower building by angling the divisions into a herringbone pattern. These herringboned cubicles work quite well—at least up to an angle of 30 degrees—but should be used only where there is no reasonable alternative.

Headrails should be fitted in cubicles to position the cows so that the dung falls into the passage. The rails will also stop cows lying too close to the front wall, from which position they sometimes find it difficult to get up. Headrails should be adjustable; the exact position will be found by trial and error but, as a guide, a typical position for a Friesian cow is 1·75 m from the passage. A breast rail, or a sleeper laid on the bed, may be used as an alternative.

Feeding Areas

These need not necessarily be covered, though a roof provides shelter for the cows, reduces foul run-off and keeps the feed dry and away from birds. The last two items are particularly important when either concentrates or a complete ration are to be fed in the mangers. The width between the feed barrier and the fence round the feeding area can range from 2·75 to 3·35 m, with 3 m as normal.

Where the feeding area forms the central section of a cubicle house, there may be either a double-sided manger about 1·5 m wide, which is filled by a forage box travelling along the cows' feeding passage, or a central drive-through, about 4·5 m wide, along which the forage box can be driven at any time without disturbing the cows. The central drive-through passage gives more flexibility for feeding different groups with different rations, and this may justify the extra cost.

Silos

Silage can be made satisfactorily in unwalled stacks, but if a high degree of waste is to be avoided, great care has to be taken over consolidation and sealing, and to reduce risk to the buckraking tractor driver the slope of the walls has to be very shallow, which means that the clamp occupies a large area in relation to its capacity. It is thus preferable to make silage in some form of container or silo, which will both reduce waste and provide safer working conditions. The two main types of silo are clamps and towers.

Clamp or bunker silos are by far the more common; they consist of a concrete floor, with walls to two sides and usually one end. Since the introduction of polythene sheet for sealing, clamps are roofed only if hay or straw is to be stored on top of the silage, or if cows are to self-feed the silage in an unfavourable climate. The walls may be constructed of reinforced mass concrete, reinforced concrete blocks, railway sleepers supported by steel stanchions, or concrete or timber panels supported either by their own struts or by the stanchions of a building.

Silo walls have to withstand not only the weight of the silage but also the thrust of consolidating tractors, the latter being exerted near the top of the wall and effecting maximum leverage. To qualify for improvement grants, silo walls have to satisfy extremely stringent Ministry of Agriculture standards, which many widely used designs no longer satisfy. Unless the walls are made of some impervious material, without joints, they should be sealed with polythene sheets to prevent the entry of air. It is important to have a safety or sighting rail 900 mm above the wall. If the silo is roofed, the eaves height should be at least 5·5 m.

Silo floors should always be laid to a fall to allow effluent to drain away, and if the silo is to be self-fed the fall must be away from the face. The acid in the effluent attacks concrete severely, and there is no complete protection against this. Painting with two coats of chlorinated rubber paint is effective but very expensive. A compromise is to use a stronger concrete mix such as C25P, which equates to the old mix of 1 cement:1·5 sand:3 aggregate.

The capacity of silos can be estimated roughly from a mean value of 800 kg per m³, but this may vary from 600 to 930 kg per m³ according to the length of chop, the dry-matter content of the crop and the degree of consolidation. The height of a clamp will depend mainly on whether or not it is to be self-fed. If it is, the walls should not exceed the maximum height to which Friesian cows can comfortably reach, that is 2·3 m. It is usual to make the walls the height of the settled silage, building up somewhat above them during making to allow for settling. In silos which will be emptied by machine, height is limited by the reach of the machinery and by the high cost of building walls above the normal height of about 2·4 m. Long narrow clamps preserve the silage better after opening but are expensive in wall. Unloading machinery and forage boxes or trailers require a minimum width of about 9 m in which to manoeuvre comfortably. On the other hand, a width of about 12 m should not be exceeded unless the silo will be emptied quickly.

Tower silos must be completely airtight. There are only two methods of construction in common use today: curved steel sheets to which a glass coating has been fused, and concrete staves bound together with high-tensile steel rings. The glass-coated steel towers require no maintenance, whereas the lining of concrete towers has to be painted periodically. Tower silos impose extremely high loads on their bases, which must be carefully installed after a thorough investigation of the soil type.

Tower silos vary in diameter from 6 to 8 m and range up to 27 m in height. Very high towers may bring objections from the planning authority. Choice of diameter and height will be a matter of balancing cost (wide, low silos are cheapest) against speed of filling and emptying (narrow, high towers fill and empty more quickly as well as consolidating better). A typical capacity requirement for dairy cows would be between 0·3 and 0·5 m³ per cow per week.

Cow Handling

Cows which have to be separated from the herd for service or veterinary attention can, in a parlour-milked herd, be diverted immediately outside the parlour exit into a holding pen. This should be large enough to hold 5 to 10 per cent of the herd and have walls, gates and railings at least 1·5 m high all round. It should be equipped with a water bowl and have facilities for tying the cows; a useful stall is made with cubicle-type divisions standing 1·22 m high and projecting 1·83 m from the wall at 760 mm centres. The cows are secured by a chain across their rumps.

For occasions when either the whole herd or a batch of heifers has to be handled, e.g. for blood sampling, it is a great convenience to have efficient facilities for marshalling and holding the cattle quickly and with the minimum of disturbance. This should not be done in the milking parlour, but it is often possible to use the fencing of the collecting yard and to bypass the parlour. It may be possible to use a space between buildings for handling, but with a large herd it is worth erecting a purpose-built handling unit (Fig. 12.7), of which the essential elements are a holding pen big enough to take a group (or at least half a group) of the cows, and a smaller, funnel-shaped forcing pen which leads into a race leading to the crush in which the handling is done. The race and crush must be strongly built in either tubular steel or pressure-treated timber, with at least four rails and a minimum height of 1·5 m. The rails should be on the inside of the posts and there should be no projecting bolt heads. The race must be so narrow that an animal has no inclination to turn round; 680 mm is suitable for adult Friesians. The crush may be proprietary or simply a yoke gate, but it must hold the animal securely, and give the veterinary surgeon good access. The working area should be roofed, well lit and equipped with a power point.

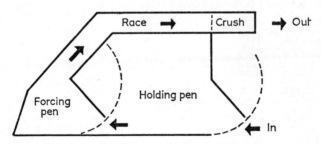

Fig. 12.7.   Cattle-handling unit

Footbaths are valuable in combating foot troubles (Chapter Seventeen). They should be at least 2·4 m long and 1·2 m wide and be capable of holding a depth of 150 mm of liquid. There should preferably be two compartments; one holding clean water to clean the cows' feet and the other containing the treatment solution. Siting in the exit passage from the parlour is convenient when the bath is in use, but at other times it can be an obstruction, and is better sited as an adjunct to the handling area.

Calving and Isolation Boxes

Calving boxes should measure at least 3·7 m square and should preferably have one dimension of not less than 4·25 m to facilitate use of calving aids. The concrete floor should fall towards the door or a drain and have a non-slip finish. Smooth floors must be avoided in calving boxes. To ease cleaning out, one wall is frequently formed by a sheeted gate, but if a door is used it must be at least 1·2 m wide and open outwards. All calving boxes should be equipped with a tying-up ring and it is a great convenience to have a crush gate (Fig. 12.8) in at least one of the boxes. This has three or four rails, is 2·3 m long, and is hinged to the wall 700 mm from a corner. A cow can be driven into the corner and the gate swung round to secure her, a chain between the wall and the gate holding her in place. Calving boxes should be sited reasonably close to the milking parlour or cowshed.

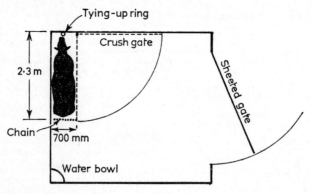

Fig. 12.8. Calving box with crush gate

All herds participating in the Brucellosis schemes are required to have at least one isolation box in which animals can be kept away from the general animal traffic of the farm. Isolation boxes should be not less than 3·7 m square, be equipped where possible with a vacuum point for milking, and have a door 1·2 m wide and opening outwards. The walls should be rendered up to a height of 1·4 m for easy cleaning, and the concrete floor should slope to a drain which can be kept separate from the general drainage system. A tying-up ring is essential and a crush gate desirable. When siting an isolation box, consideration should be given to the possible removal of corpses.

Young Stock

Three essentials of good calf housing are a dry bed, good ventilation and freedom from draughts. If these are provided and nutrition is good, the calf can tolerate low temperatures. Ventilation by fan, termed 'controlled environment,' can be effective and in many conversions of existing buildings provides the only satisfactory answer, but in new buildings it should be possible to provide the right conditions with natural ventilation. Natural ventilation relies on 'stack effect', which means that air warmed by the calves' bodies rises by convection and escapes through the outlets, drawing fresh air in through the inlets, which are above the level of the calves. Downdraughts on to the calves are prevented by providing simple covers over part of the pens. The requirements for natural ventilation are: an air space of 6 m³ per calf, a height difference between outlet and inlet of from 1·5 to 2·5 m and (in a fully stocked house) an inlet area of 0·045 m² per calf and an outlet area of 0·04 m² per calf.

Calves on removal from their dams may be housed in either individual or group pens. Individual pens have the advantage of reducing the spread of disease and eliminating navel sucking; for Friesian calves, the minimum size should be: up to 4 weeks, 1·5 m × 750 mm; up to 8 weeks, 1·8 × 1 m.

To ensure a dry bed, the pen floors should be either of wooden slats or of insulated concrete with a slope of 1 in 20; drainage should be to floor channels which quickly remove all moisture from the house. Pens should either be easily dismantled or easily cleaned *in situ*.

Group penning of young calves is common practice, and satisfactory provided that management is good. A floor space of 1·1 m² per calf is required up to 8 weeks, and 1·5 m² up to 12 weeks. Each calf needs 350 mm of trough space unless an automatic feeder is used. After weaning, group pens in strawyards are usual. A good design provides a straw-bedded area, with a concrete area next to the trough which is kept clean. This reduces straw use and helps to keep the calves' feet in good condition. Autumn-born calves in their first winter require about 3·0 m² of floor space and 450 mm of trough space. In their second winter they require 4·0 to 5·0 m² of floor space and 500 mm of trough space.

Heifers may be kept in cubicles from birth right through their lives, and this practice can save much work and bedding, as well as accustoming the animals early to their lifelong form of housing. Suitable cubicle sizes for animals in their first and second winters are, respectively: 1·5 m × 700 mm, and 2·0 × 1·0 m, but some trial and error and adjustment may be necessary.

There are a few successful units in which the cubicles for second-winter heifers are not roofed over. Provided that the beds are porous and the site well sheltered, the stock do not appear to suffer, though they lie dirty. Rainfall on the fouled concrete areas presents a problem with foul run-off.

Bull Pens

Bulls are dangerous animals, and their accommodation must be particularly solidly built, and designed so as to give the bull reasonable exercise space and, if possible, a view of the cows. Bull pens normally consist of a covered box and an outside run. The box should have an area of about 17 m², with 225 mm thick brick or block walls rendered up to a height of 1·4 m, and an insulated floor. The door from the box to the run must be 1·2 m wide and at least 2·1 m high. There should be a water bowl protected by a concrete plinth, and a manger which can be filled from outside the box.

The outside run should have an area of about 33 m² with walls at least 1·5 m high, part of which should consist of rails above a height of 1·14 m to permit the bull to see out. The rails must be galvanized, with a minimum internal diameter of 38 mm, and not more than 150 mm apart; they must be supported by steel stanchions well concreted into the ground at centres not exceeding 1·8 m, and not bedded in the concrete blocks. A service pen 1·07 m wide eases management.

It should be possible to contain the bull in both the box and the outside run while the other part is cleaned out. This may be achieved by the use of slip rails across the doorway of the box, or by a yoke fitted to the manger. As a safety precaution, a refuge should be formed in a corner, using steel joists.

General Layout

Integration of the separate buildings into a coherent overall layout is as important as their individual design. How this will be done will vary with the individual site, but there are certain principles which have general application. It should (at least in herds of over 90 cows) be possible to divide the herd into two or more groups for winter feeding, and movement of cows between their living quarters and the parlour should be easy. It should be possible to put out feed and scrape slurry while the cows are in the collecting yard. The slurry store should be convenient to the living and collecting areas, and slurry scraping should involve the minimum of turning. The layout should be as compact as possible without being cramped, and uncovered fouled concrete should be kept to a minimum to reduce foul run-off. Provision should always be made for future expansion. Feeding methods have a great influence on layout: forage boxes need plenty of turning space, whereas self-feed silage systems will be more compact, although ample room must be left for tipping trailers during silage making. The individual site will affect layout by the way the levels run, or perhaps by the presence of an existing road close to which it will clearly be sensible to build the dairy.

Fig. 12.9 shows a typical layout for a herd of 130 cows, milked in a 10/10 herringbone and split into two groups for winter feeding. A central drive-through passage is provided for feeding, although the silos have been sited so that self-feeding can if required be practised without difficulty, and in two groups. The cows are under cover all the time. If this were designed mainly as a self-feeding layout, the passage giving the cows access to the collecting yard would be omitted, and the silos and the access area covered by continuing the roof of the cubicle house at a higher level.

The most common faults of layout relate to space; either there is not enough and the layout is cramped, or there is waste, with expensive and unnecessary areas of concrete. Another very common fault is to build without thought for expansion.

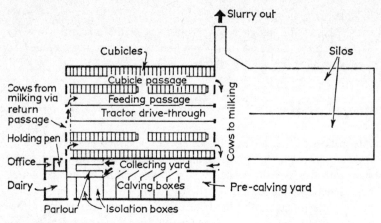

Fig. 12.9. Typical layout of a dairy unit

Planning Regulations and Appearance

In general, and outside certain areas, farm buildings only require planning permission if: either they are within 3·0 km of the perimeter of an aerodrome; or they are within 25·0 m of a classified road; or they obstruct the view on a road; or if the new building, aggregated with other buildings erected within 90·0 m within any period of 2 years, totals 465 m² in area. If there is any doubt, reference should be made to the planning office of the District Council. The processing of a planning application takes time; a decision is frequently made within 2 months, but it is safer to allow 3 months. Fees are payable on planning applications relating to buildings over 465 m² in area.

Modern dairy buildings are often large and, quite apart from satisfying the planning authority, it is important to see that they have a pleasant appearance, or at least offend the eye as little as possible. The impact of a large building on the landscape can be lessened most effectively by careful choice of site, but in practice the siting of dairy buildings is usually dictated within narrow limits by economic and functional considerations. In general, improving the appearance of a building tends to cost money, but there are many, not necessarily expensive ways in which an acceptable appearance can be achieved. Eaves height will usually be dictated by function, but there is often scope for matching roof pitch to that of nearby buildings. Cladding materials also can often be matched to the existing buildings without great expense. The outline of a new building can be softened by the planting of a few well placed trees.

Further Reading

Cost of buildings handbook, 1982, Ministry of Agriculture, Fisheries and Food, London

Dairy cattle facilities, Farm Buildings Topics No. 13, 1981, East of Scotland Agricultural College, Edinburgh

Design of buildings and structures for agriculture, B.S.I., B.S. 5502, 1981, Section 2.2, Livestock Buildings

Harrison, E. R., *Planning dairy units*, 1979, Farm Buildings Information Centre, N.A.C., Stoneleigh, Warwicks

Mitchell, C. D., *Calf housing handbook*, 1976, Farm Buildings Information Centre, N.A.C., Stoneleigh, Warwicks

Wight, H. and Clark, J., *Farm buildings cost guide*, 1982, Scottish Farm Buildings Investigation Unit, Craibstone, Aberdeen

CHAPTER THIRTEEN

Slurry

Composition of Slurry—The Legal Position—Moving Slurry—Slurry Storage—Short-term Storage—Long-term Storage—Earthwalled Compounds—Foul Run-off—Separators—Organic Irrigation

---

Slurry consists of dung and urine which, while they may contain some waste food and bedding, have been dropped on concrete rather than on a littered surface. The quantity produced daily by a cow is approximately 8 per cent of its bodyweight, the exact amount depending on the ration being offered. A Friesian cow will void about 0.04 m^3 (40 l) of dung and urine per day, or about 7 m^3 (7,000 l) in a winter of 180 days. The widespread adoption of cubicle housing has greatly increased the slurry problem, but it also arises to a lesser degree in other housing systems.

Composition of Slurry

Undiluted slurry contains approximately 87 per cent water, but this value varies widely with the type of food. The content of nitrogen, phosphorus and potassium—N, P and K—also varies widely according to the proportions of dung and urine which the slurry contains; as the proportion of urine increases, the content of K increases, and the ratio of K to N widens. On average, 10 m^3 (10,000 l) of undiluted cow slurry will contain 25 kg N, 10 kg P_2O_5 and 45 kg K_2O in the form of nutrients available to the crops in the first year after a spring dressing. Further nutrients may become available subsequently, but this is difficult to measure. The available N in slurry spread in autumn is almost completely lost by leaching during the winter. Grassland can respond to applications of slurry in a similar way as to dressings of a compound NPK fertilizer, the response being governed by the ratio of dung to urine in the slurry. On many intensive farms, the nutrients in the slurry have about one-third to half the value of the purchased fertilizer and the potential value of this by-product, if properly stored and applied, should never be underestimated.

Slurry should where possible be spread on grassland which is to be used for conservation rather than grazing, and to prevent pollution and undue build-up of fertilizer nutrients in the soil, single applications should not exceed 55 m³ (55,000 l) per ha and the interval between applications should be at least 2 weeks.

The Legal Position

It is an offence to cause, or knowingly permit any noxious or polluting matter to enter any river, stream or inland water, including any lake or pond which discharges into such a watercourse. The Acts governing such matters set out a number of statutory defences to a charge of pollution, one of which is that the act causing the pollution was in accordance with good agricultural practice as set out in codes of practice drawn up by the Ministry of Agriculture. These codes lay down, among other things, maximum rates of slurry application, stressing their relationship to the nutrient requirements of the crop.

The stringency with which pollution control is applied varies considerably in different parts of the country, but it must be expected to become increasingly severe and to have an increasing effect on the design and management of dairy units.

Smell can be a serious problem near towns and villages, and must be kept in mind when planning the storage and spreading of slurry in such vicinities. In general, slurry smells more strongly the longer it is stored and the more water it contains.

Moving Slurry

Slurry is normally moved from the concreted cubicle passages and yards to the store by tractor scrapers. These are usually rear-mounted, with a renewable rubber blade, and should preferably be reversible. Steel blades should be avoided. Passages may also be scraped by permanently installed mechanical scrapers, of which the most common type comprises a fixed or folding blade which is drawn by a chain operated by an electric motor. The reliability of these scrapers has improved and they are now gaining in popularity.

With slatted passages the cows tread the dung between the gaps in the slats into the cellar or channel below. No scraping is needed, and the cows keep exceptionally clean. The cellar may, if the water table of the soil permits, be made deep enough to hold several months' slurry, but deep cellars require strongly built and expensive walls, and a more economical alternative may be to use a shallow cellar or channel designed to allow the slurry to flow to one emptying point. The minimum depth of such a channel should be 3 per cent of its length plus 400 mm, and to prevent the liquids from draining away separately from the solids, the floor of the channel should be laid level and should incorporate weirs to retain a depth of about 150 mm of liquid.

Slurry will flow by gravity provided that it does not contain too much waste food or litter; flow characteristics depend on the dry-matter content, which in turn is determined by the degree of dilution and on the feeding. If there is space, slurry will flow between earth banks, but it is more usually contained in concrete block channels.

Slurry Storage

When planning a new dairy unit or modifying an existing one, it is essential to decide at an early stage how the slurry is to be stored and handled.

Exceptionally free-draining soils which will carry machinery without damage throughout the winter permit the simplest and cheapest form of slurry handling, which is to scrape it direct into a spreader for daily disposal. Use may sometimes be made of existing slopes to stand the spreader at a lower level, but more usually a concrete ramp is needed. This should have substantial kerbs and safety rails, and to avoid wheel-slip must not be steeper than 1 in 8.

Short-term Storage

To avoid spreading at weekends and in exceptionally wet weather, it is preferable to have at least some storage capacity. For semi-solid slurry (i.e. slurry which contains enough waste food and litter for it to be handled with a fork) a small dungstead will suffice, comprising a concrete base and low walls, which may be of sleepers or even earth. Slurry is loaded into spreaders with a tractor and foreloader.

More liquid slurry will be stored at greater depth in either pits or above-ground stores. The former may consist of a pit lined with reinforced concrete blocks, or simply a hole in the ground, and will be emptied by tanker. Such stores must be planned with careful regard to the water table, i.e. the level to which soil water settles in winter. Above-ground stores usually consist of tower-silo sections made of concrete or glass-lined steel. The slurry is scraped into a reception pit, from which it is transferred by pump or auger into the store. Such stores are expensive, but may be the only possibility where the water table is high.

Short-term storage is attractive in its low initial cost and because it needs little space. It also avoids either the heavy cost of emptying a large slurry store by contractor, or a heavy workload imposed on farm staff at what may be a busy time. Its disadvantages include the constant demand on labour for spreading, the damage to swards and soil structure and the waste of fertilizer nutrients due to spreading at the time of year when least use can be made of them.

Long-term Storage

Long-term storage usually involves containing the whole winter's slurry for disposal during the following spring and summer. The two main types of store are tower-silo sections and earth-walled compounds (Fig. 13.1).

The tower silo store is identical, except in size, to that used for short-term storage. The largest currently available has a diameter of 25 m, a height of 6·1 m and a capacity of about 2,800 m^3 (2·8 million l). Long-term storage in these large towers requires careful management; frequent recirculation is essential if serious crusting is to be avoided, and if the pump or auger is to work satisfactorily some dilution with water is necessary and the minimum of waste food and litter must be admitted. This means that a dungstead will also be needed. Tower stores are compact and tidy and may offer the only solution where the water authority will not permit the use of an earth store; they can also, if properly operated, keep the slurry in ideal condition for spreading on grassland, and enable it to be applied at the optimum time. Tower stores are, however, expensive both in first cost and in maintenance of the machinery, and they make considerable demands on management.

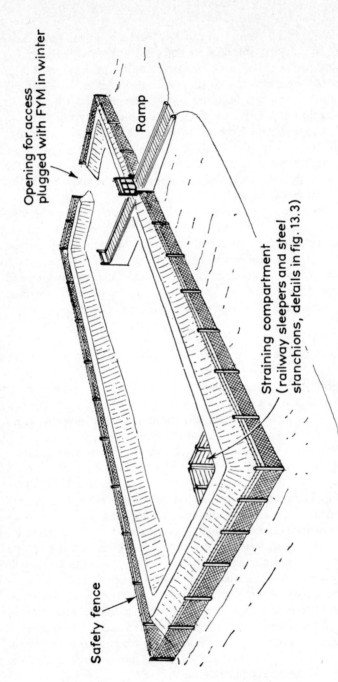

Opening for access
plugged with FYM in winter

Ramp

Safety fence

Straining compartment
(railway sleepers and steel
stanchions, details in fig. 13.3)

Fig. 13.1. Earth-walled slurry compound (FYM = farmyard manure)

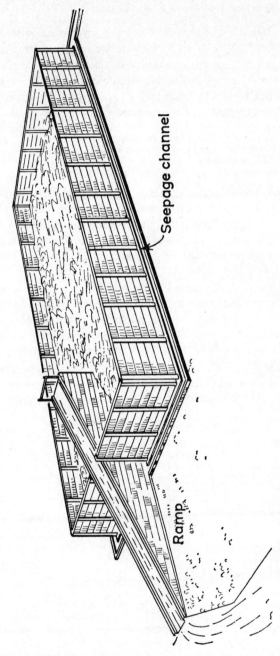

Seepage channel

Ramp

Fig. 13.2. Sleeper-walled slurry compound

Another type of above-ground store is formed of either railway sleepers (Fig. 13.2) or concrete sections with vertical supports. The slurry is scraped into these compounds up a ramp, and the liquid seeps out through gaps in the walls for collection and disposal. The solids dry out by the summer and a section of wall is removed to permit farm equipment to enter for emptying.

Earth-walled Compounds

These are often incorrectly called 'lagoons', which they are not. A lagoon is a wide, shallow store in which highly diluted slurry is stored and in which the organic matter is slowly digested by aerobic bacteria. This system, which was evolved in the United States, does not work satisfactorily in our cooler climate and is unacceptably wasteful of fertilizer value.

An earth-walled compound is formed by excavating an area of earth and using the excavated material to form walls. The depth of excavation will vary with the site, but typically will be about 1·2 m, with the walls being built about 1·2 m above existing ground level, giving an overall depth of 2·4 m. The walls must be carefully designed and thoroughly consolidated during construction; there have been many instances of walls leaking or even giving way because they were built too steep and narrow and were inadequately consolidated.

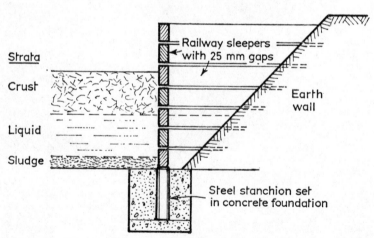

Fig. 13.3. Cross-section of straining compartment, showing stratification of slurry

When slurry is stored to any depth it has a strong tendency to stratify (Fig. 13.3), i.e. for the fibrous material to rise to the surface, forming a steadily thickening crust, and for a small amount of sandy sludge to sink to the bottom, leaving a gradually clearing liquid layer between them. Choice of method of emptying the compounds depends principally on whether it is decided to handle this liquid layer separately or to try to mix it with the solids again, thus restoring the slurry to its semi-liquid state. If the latter course is taken, emptying is by pump or tanker, and considerable dilution is required as well as the prolonged use of propeller-type agitators. Compounds for this semi-liquid system of emptying are deep and narrow, with a steep fall on the floor towards the emptying point. This semi-liquid slurry is very suitable for spreading on grassland, but emptying by pump or tanker is unlikely to succeed in removing all the solids; the tendency is for the liquid layer only to be removed, leaving the solids behind. Use of a digger is thus likely to be necessary in addition.

Where the liquid layer is to be disposed of separately, it is essential to keep out all drainage and foul run-off. Some form of straining compartment is usually provided to facilitate removal of the liquid. This may be a simple tube of welded steel mesh and nylon netting, but a more substantial barrier can be formed of railway sleepers slotted into steel columns and wedged about 25 mm apart (Fig. 13.3). Such barriers hold back the solids but allow the liquid to pass through into the straining compartment, from which it may be drawn off in the spring by pump or tanker or led off through a pipe to a secondary storage compound, which can also be useful for storing parlour washings and foul run-off.

After removal of the liquid, the remaining solids will be of a consistency which is easily handled by conventional spreaders. If these are to be filled by farm equipment, which will usually consist of a tractor and foreloader, a concrete floor to the compound is essential on most soil types, and the compound will preferably be wide and shallow. Probably the most usual contractor's machine for emptying is the swivelling-boom digger. The largest version of this machine which is generally available has a horizontal reach of about 8 m; it will normally operate from the top of the banks, which should be at least 3 m wide to accommodate it, and will fill spreaders standing outside the walls. Draglines have a reach of about 12 m, but are slow and clumsy in operation and suited only to emergency use.

When designing a compound for emptying by contractor's machine, it is essential either to ensure that the whole of the contents can be reached from the banks or to make provision for a smaller machine to go down into the compound to push the last of the slurry within reach of the digger. Hire of a digger and a team of spreaders is costly, and a good traffic flow must be ensured.

The cubic capacity of a compound can be calculated by multiplying the daily slurry production of the herd (0·04 m³ per cow per day) by the number of days in a normal winter-housing period, an allowance being added for litter if much straw is used. Further allowance should be made for the direct rainfall on the compound, and finally a safety margin of about 600 mm should be added on to the height of the banks.

Soil pores are quickly sealed by slurry, and efforts to allow even foul run-off to soak away into the soil are unlikely to be successful for long on any soil type. Such practices are in any event unlikely to be permitted by the local water authority, which should be consulted at an early stage in the planning of any slurry disposal system.

The crust which forms on slurry offers a deceptively firm appearance, and can be a lethal trap for children and animals. Safety regulations with regard to the fencing of compounds are rightly very stringent, a typical specification being a fence 1·4 m high composed of chain link surmounted by two strands of barbed wire. The cost of such a fence is a considerable item, which must be allowed for when comparing the cost of various storage systems.

Foul Run-off

One of the most difficult problems on many farms is the disposal of the water which runs off open concrete which has been fouled by cattle. In the past, this liquid went into a ditch and thence to the nearest watercourse, but this is now illegal. The worst problems arise where there is an uncovered self-feed silo, with perhaps an uncovered feeding area and collecting yard in addition. For a herd of 100 cows, this area can total 800 m², and 25 mm of rain produces 20,000 litres of liquid for disposal. Since winter rainfall in many parts of the country is 500 mm or more, the size of the problem can readily be seen.

The first step in trying to deal with this problem is to ensure that all rainwater falling on roofs and clean concrete goes straight into a ditch. The contaminated water, or foul run-off, can be spread on the land as it is produced, or stored for later spreading, or purified. Purification, by means of settling ponds and sedimentation ditches, is the cheapest method of disposal, but is of dubious effectiveness. It will make some reduction in the biological oxygen demand (BOD), which is a measure of the polluting effect of a substance, but the degree of the reduction will depend on the length of ditch, the regularity with which it is cleaned out and the type of effluent which is put into it. It is useless to attempt to purify silage effluent or neat slurry. Purification can be taken a stage further with the use of towers and aeration devices, but efforts to purify foul run-off, however elaborate, are unlikely to be successful in bringing the BOD down to a level acceptable to a water authority, and the more complex systems cannot be economically justified.

If slurry is stored anaerobically before spreading on the land, some pathogenic organisms and worms may survive in significant numbers for several months. Probably the most important is *Salmonella*, and in addition *Brucella*, *Leptospira* and *Mycobacterium* and many viruses can survive for long periods. Aerobic treatment of the slurry reduces the survival time of the pathogens, but many may still be present unless the temperature of the slurry is raised or the aerobic treatment prolonged.

The dissemination of the organisms depends on the spreading technique. Rainguns cause a high production of an aerosol type of spray risk and with soil injection there is little or none. Infection can occur if cattle drink from a watercourse downstream from land drains which are contaminated with slurry. To avoid this risk, slurry should not be applied when it may enter a watercourse, e.g. when the soil content is at field capacity or when the ground is frozen hard.

Other precautions to reduce possible health hazards are to store all slurry for 3 to 6 months, not to use rainguns, especially in windy weather, and to avoid grazing a field for at least 21 days after an application of slurry. It is always preferable to use grass from fields which have had a dressing of slurry for conservation rather than grazing. The risk of spreading disease from slurry should not be exaggerated, but the potential danger must always be appreciated and the necessary precautions taken.

Separators

There are several different types of machine available for mechanically separating the fibre in the slurry from the liquid. They work on various principles, including the vibrating screen, rotary screen and roller, gravity and centrifuge. They remove nearly all the fibre and leave two easily handled materials: the solids, which are eaily stacked and spread, and the liquid, which has little tendency to form a crust and which can be pumped like water. These machines are expensive in first cost and in maintenance, but there are some circumstances in which the extra efficiency of separation which they provide is justified.

Organic Irrigation

In this system the slurry is kept as free as possible from waste food and litter, and is scraped into a pit in which it is diluted with at least double its volume of water and kept stirred by a mechanical agitator. The size of the pit is limited by the reach of the agitator, and emptying has to be frequent. The diluted slurry is usually drawn through a chopper or conditioner and then pumped out through pipes to a raingun, which may move automatically, in the field. The advantage of the system is that the slurry is pumped rather than mechanically handled, and is disposed of as it is produced. Its disadvantages include the unpleasant task of moving the pipes where an automatic raingun is not used, the risk of mechanical breakdowns and of sward damage if the guns are not moved frequently enough, and the possibility of slurry running off the surface or penetrating to the land drains and thence to watercourses. The system is difficult to operate in frosty weather, and can cause nuisance near houses through the small particles which are thrown high into the air. Owing to the limitations of the pump, slurry tends to be applied to the nearer fields, which are usually those which need its fertilizer value least.

Further Reading

Farm waste management—profitable utilization of livestock manures, Booklet 2081, 1982, Ministry of Agriculture, Fisheries and Food, London

Farm waste management—general information, Booklet 2077, 1982, Ministry of Agriculture, Fisheries and Food, London

Robertson, A. M., *Farm wastes handbook*, 1977, Scottish Farm Buildings Investigation Unit, Craibstone, Bucksburn, Aberdeen

The storage of farm manures and slurries, Booklet 2273, 1980, Ministry of Agriculture, Fisheries and Food, London

Weller, J. B. and Willetts, S. L., *Farm wastes management*, 1977, Crosby Lockwood Staples, London

CHAPTER FOURTEEN

Herd Management and Records

Identification—Grouping of Cows—Records—Health Records—Breeding Records—Milk Recording—Herd and Group Records—Body-condition Scoring—Labour—Bonus and Incentive Schemes—Relief and Contract Milking

The profitability of milk production, as of any other industry, depends on good management. This involves good identification of the animals, a detailed and up-to-date knowledge of what is happening in the business, and an efficient and contented staff.

Identification

Management starts with knowing which cow is which. Unless each animal can be immediately identified it is impossible to tell when she calved, whether or not she is due for service, or the amount of concentrates which she should receive. In small herds, the person who does the milking will know the cows individually without the help of any special identification mark, but even in such herds there will be holidays or times of sickness when somebody else has to milk, and a positive means of cow identification is a great help. In larger herds, a positive and permanent means of identification is absolutely essential.

Freeze-branding is a permanent, and the most generally adopted, method of identification. The hide is clipped, sprayed with alcohol, and a brand applied which has been chilled to an extremely low temperature. This temperature kills the pigment cells of the hair which, when it grows again, is white. Freeze-branding is a highly effective way of marking dark-haired cows such as the Friesian, but is less effective with the Channel Island and Ayrshire breeds. In theory heifers can be branded at any age, but if the animals are too young the brands may distort. A good compromise is to brand heifers at the beginning of the winter during which they will be served;

they will then be readily identifiable during the service period, making record keeping much easier. Branding may be either on the back of the leg, where it can most easily be seen from the pit of a low-level parlour, or on the rump, where it can be seen more easily at other times and as the cows walk into the parlour. Branding is easier on the back of the leg, since there is less movement of the animal. Freeze-branding is normally done by contractors.

Other means of identification include neck chains, coloured and numbered plastic collars, anklets and eartags with numbers. Alternatively, transponders can be fitted which transmit electronically the number of the cow.

Since it is impracticable to freeze-brand a cow with a name, it is usual to give each cow a separate herd number. In some circumstances, it is desirable to keep the same group of numbers, e.g. 1 to 50, for heifers entering the herd and taking over the numbers of the old cows which they replace, but in general it is better to number the animals consecutively, so that the age of a cow can be judged roughly from her number. In some herds, it may be an advantage to precede each heifer's number with a letter indicating the year: e.g. A1, A2, etc. for year 1; B1, B2, etc. for year 2.

There are occasions when temporary marking of cows is required for management purposes, for example to indicate a cow with mastitis whose milk has to be discarded. Such temporary marking can conveniently be done with a patch of paint or dye from a paint stick or an aerosol can, which will last for a few days and can be renewed if necessary. The level of concentrates which a cow should receive in the parlour can be indicated by use of a colour code using different coloured sticky tape round the tail; these tapes can be renewed or changed as required.

There is a statutory requirement that every animal shall have an eartag showing both the herd number (which is allocated to every herd by the Ministry of Agriculture) and the individual number of the cow, which is decided by the farmer.

Grouping of Cows

Most British herds of more than about 100 cows are divided into two or more groups for winter management. The smaller the group, the nearer one can come to the ideal of treating each cow as an individual. However, a large number of small groups complicates management and can delay milking as fresh groups of cows are brought to the collecting yard. A sensible compromise is to separate the dry cows and then divide the milking herd into groups of between 50 and 80 cows. This enables different rations to be offered to the various groups outside the parlour and also allows the higher yielding cows to be milked first in the morning and last at night.

Apart from increased efficiency of feeding, it is believed that cows have a preference for living in comparatively small groups, the theory being that in groups of above about 90 they cannot identify their neighbours at the manger and thus are uncertain of their respective places in the social or 'pecking' order. This is said to lead to stress: whether this is really so remains unproved.

Another matter for debate is whether it is better to group cows according to their milk yield, and move them between groups periodically, or to group them simply by calving date, with no switching between groups. The former system is the more logically efficient from the feeding point of view. It does, however, impose a dual stress on the cow being moved down a group: a change of group-mates and of feed at the same time.

Records

Even the smallest herd cannot be run efficiently without some simple records; on the other hand it is a mistake to keep records for their own sake. A record should never be kept unless there is some definite use for it, and unless there is a full intention of keeping it properly.

The things which it is essential to know when managing a herd are quite simple: the state of the cows' health; whether the cows are conceiving at the right time; the level of milk yield; and whether the correct amounts of concentrates are being given. These are the few factors which it is useful to monitor regularly throughout the year; there are other equally important items, such as stocking rate and labour efficiency, which are more suitable for periodic review than for routine monitoring.

The use of micro-computers for the keeping of such records is increasing rapidly. Programs or 'software' are available for both general purposes such as the keeping of accounts and the payroll, and for specialized dairy records. A typical financial program will show the up-to-date stock position of feeding stuffs and fertilizer, the bank situation, and information for the V.A.T. return. It will produce also a monthly financial summary with a comparison with the figures from the previous year and with a budget. At the end of the year it will produce a profit and loss account with enterprise gross margins compared with budgets. The payroll program will calculate wages, including overtime and deductions, and produce payslips.

The specialized programs are equally varied, but usually include an individual cow record giving cumulative milk yields, margin over concentrates, and breeding and veterinary information. From the records, a weekly action list may be produced which lists the cows due for service and drying off. The performance of selected groups of cows compared with standard lactation graphs is also usually monitored, and there are normally facilities for calculating the amount of feed which each cow should receive, and for summarizing the herd health situation and breeding performance.

Health Records

A record should be kept of every veterinary treatment of every cow, in particular udder infusions for mastitis and dry-cow therapy. Keeping a diary in the dairy office takes little time and trouble, and the veterinary surgeon may well be pleased to make the entries himself for his own benefit at subsequent visits. A more elaborate system is to have a card for each cow on which a complete health history is recorded. Both systems can work well, but a diary is more likely to be kept up, and is more useful for reminders of such items as ovary massages and for showing up herd trends such as outbreaks of foot trouble or mastitis. The individual cards, on the other hand, give a concise picture of each cow's history which the diary cannot match, and can help in pinpointing cows which should be considered for culling.

All farmers are required by law to keep a Movement of Animals book in which all movements of livestock to and from the farm are registered.

Breeding Records

The first and most important thing to be recorded about any cow is the date on which she calves; this information must be written down in some permanent record, no matter how elaborate a system of visual aids may be used. It is also highly desirable to keep a formal record of all services, showing the date, the breed and name of the bull, the identification of the cow and the number of the service. This information could, in an artificially inseminated herd, be obtained by looking through the insemination certificates, but a separate record is well worth keeping.

To ensure that a cow is served at the right time, it is necessary to know the date of calving; whether the cow has been seen on heat (so that, if she has not been seen, routine veterinary attention can be given at, say, 5 weeks after calving); the dates of heat periods; and whether the animal is ready for service, i.e. whether the minimum time has elapsed between calving and service. It is also useful to be warned when 3 weeks have elapsed since service, so that a particularly careful watch can be kept for signs of heat. All this information can be presented graphically by various visual-aid systems, which are an invaluable aid to herd management. They usually take the form of a calendar, either circular or rectangular, on which numbered pins or pegs represent the individual cows. The circular type probably gives a clearer picture of a herd with up to about 100 cows, but above this number the circle tends to become overcrowded and the rectangular type is to be preferred. A familiar type of circular cow calendar is

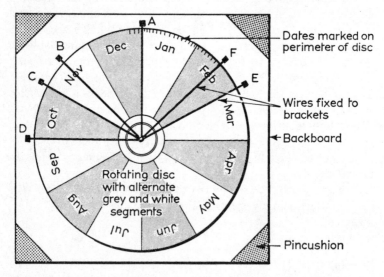

Fig. 14.1. Circular cow calendar
Wires: A = today's date and calving date; B = serve cows (6 weeks after calving); C = serve heifers (8 weeks after calving); D = service line (12 weeks after calving); E = drying-off (8 weeks before calving); F = steaming-up (6 weeks before calving)

shown in Fig. 14.1. For each cow there are four large indicator pins of different colours, on which her number is written. A suggested colour code is: green for calved cows; blue for cows which have been seen on heat; red for served cows; and yellow for dry cows. The disc is rotated daily so that today's date is beneath wire A. When a cow calves, her green pin is inserted under wire A. When she is seen on heat, her blue pin is substituted for the green one. When she is served, her blue pin is removed and her red pin inserted under the service wire. If she returns, the red pin will be replaced under the service wire; if she holds to service, the pin remains in place until it reaches the drying-off wire, when it is replaced with a yellow pin.

These visual aids show up vividly, in a way which no written record can, any individual abnormality, e.g. a cow which has still not been served 12 weeks after calving, or which is overdue for drying-off. They also show clearly the herd's calving pattern. Much of this breeding information is provided by milk-recording schemes, but a visual aid on the dairy office wall is a daily reminder as well as a record.

Milk Recording

Milk recording involves the regular measurement and recording of the milk output of the individual animals in a herd. These records are of great assistance in selecting cows for breeding or for culling, and as a guide to feeding.

Milk recording can be done privately, but membership of an organized scheme has the great advantage of imposing strict standards and a regular discipline, as well as ensuring credibility. Recording schemes are generally organized by the Milk Marketing Boards, and vary according to the country. In England and Wales there are three types of milk recording service available, in Scotland two main types and in Northern Ireland a single scheme. The percentage of all cows which are officially recorded is 44 per cent in England and Wales, 45 per cent in Scotland and 8 per cent in Northern Ireland.

In all the official schemes the milk is either weighed in buckets or jars, or measured in graduated jars or through milk meters. An official milk recorder visits the farm for two successive milkings every month, records the milk yields and takes samples for the determination of the fat and protein contents. The recorder also records details of calvings, services, drying-off and cows leaving the herd. Following the visit, the milk producer receives a summary of the results which, depending on the particular scheme involved, may include for each cow the milk yield and fat and protein percentages on the day of the visit, and a herd average for all three statistics for that day. It may also show the calving date and lactation number for each cow, with a cumulative lactation yield and rolling averages of fat and protein percentages. It may also contain forecasts of calving date and show the peak service period.

Following completion of every lactation, the producer receives a record card for each cow which gives details of her breeding, birth date, and lifetime performance, and is the principal source of information when evaluating the cows for breeding purposes. In the main scheme in England and Wales the card includes an index number which gives the estimated value of the animal in relation to the rest of the herd, judged on milk yield and quality in the latest lactation and adjusted for age. The quality of the herd on the same basis, relative to other recorded herds, is shown as well. The index is not intended to form the sole basis of culling decisions, but simply as a quick indication of relative merit based on figures only. An annual summary shows the performance of every cow and heifer completing a lactation in the recording year, averages for the herd, and the calving index, i.e. the number of days between successive calvings.

Herd and Group Records

These records link milk production and feed use. There are a number of costing schemes available in which at the end of each month the requisite data are sent in by or collected from the farmer for processing by a computer. The data required from the farmer include cow numbers, milk sales and feed used, and the statement returned to him shows the rolling milk yield, the rate of concentrate use, and the rolling margin over purchased feed both for the herd and per cow. These results may be shown in comparison either with the results of the same herd for the previous year or with a 'league table' composed of the other herds in the same scheme, the herds being identified only by numbers. Such records are essentially retrospective, and while they are valuable in maintaining a check on efficiency and progress, they are of limited help in making the immediate, short-term feeding decisions which are so vitally important, particularly in the peak period of lactation.

If recording schemes are to be a real help to immediate management, they must contain an element of forecasting. One of several such schemes is provided by the Milk Marketing Board of England and Wales. Briefly, the data from the monthly milk records, together with information on calving date, age, previous yield and date of next calving, are used to produce by computer a prediction of the weekly milk yield of each cow for the next 6 weeks. The Board's adviser visits the farm, corrects the data for any unpredicted changes which have taken place since the last milk recording, and produces from them a graph of predicted weekly milk yields for the whole herd for the next 6 weeks. The farmer plots the actual weekly milk production on the same graph in a different colour, and the relationship between the two lines provides a sensitive indication of how successfully production targets are being met. This scheme is based on a detailed study of many thousands of lactation curves to produce a series of standard curves for cows calving in the different months of the year.

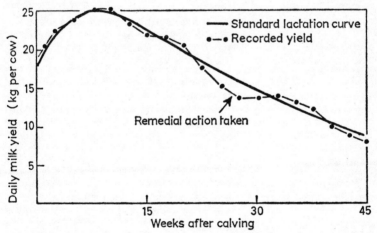

Fig. 14.2. Daily milk yield of a group of cows, plotted against a standard lactation curve

As an alternative to such computer-controlled schemes, performance can be simply monitored in the farm office, using only a single standard lactation graph and simple arithmetic or a pocket calculator. A useful method, particularly in herds with a seasonal calving pattern, is to select, say, 10 to 12 cows calving within a short period during the peak calving season, e.g. from 10 to 31 October, record their yields weekly and plot their average on a graph against a standard lactation curve (Fig. 14.2). A deviation of the recorded milk graph below the standard curve will immediately indicate some deficiency in the feeding and management, and remedial steps can be taken at once. During peak lactation, a week of incorrect feeding can damage a whole lactation.

Table 14.1. Body-condition scores for Friesian cows

Score	Condition	Loin	Tailhead
0	Emaciated	Individual transverse processes are 'sharp' to the touch; no fat cover on top	Deep cavity around tailhead and under tail; skin tight against bone
1	Poor	Ends of transverse processes less sharp, but top can be felt	Cavity around tailhead; skin slacker, but no fat
2	Moderate	Ends of transverse processes have cover and feel 'rounded'; top felt with pressure	Cavity around tailhead filling with fatty tissue; skin supple
3	Good	Ends of transverse processes felt only with pressure; thick layer on top	Cavity filled with fatty tissue; skin round and smooth
4	Fat	Ends of transverse processes cannot be felt, even with firm pressure	Cavity overfilled with fatty tissue; patches of fatty tissue under skin
5	Grossly fat	Folds of fatty tissue over transverse processes; bones cannot be felt	Tailhead buried in fatty tissue; pelvis cannot be felt

(Based on data of Mulvany, P. M., in *Dairy cow condition scoring*)

Body-condition Scoring

Condition scoring is a technique for assessing the reserves of sub-cutaneous fat in the cow's body, and the results can be used as an aid to the feeding and management of the dairy herd. The assessment of the fat is made in the loin area (Fig. 14.3) and the area around the tailhead. The body condition is scored on a numerical scale from 0, which represents emaciation, to 5, which indicates gross over-fatness (Table 14.1). The technique can be operated as either a visual assessment or by handling, and, with experience, can give reliable and consistent results.

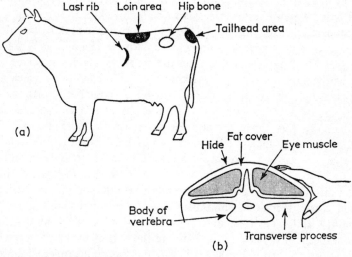

Fig. 14.3. Body-condition scoring
Key: (*a*) areas of the cow's body for assessing body condition—loin area and tailhead area; (*b*) cross-section of loin area of cow, indicating lumbar vertebrae and the position of fingers and thumb when scoring
(Based on *Condition scoring—dairy cows*)

The scores are used as follows. At calving a cow must be in good condition so that the animal can express her potential yield, and a score of 3·0 to 3·5 is ideal. At the time of service the cows should be at score 2·0 to 2·5, so that there is an increased chance of the animals holding to the bull. When drying off a score of 3·0 to 3·5 is again required, so that the animals are in good condition for the next lactation. Scores of 0·5 to 1·5 can result in lowered milk yields and a poor rate of conception, whereas scores of 4 and over indicate overfat animals which are often low yielders.

The scoring of the cows should preferably be made by the person who regularly handles them. Animals should be recorded at calving, and at 8, 16 and 30 to 35 weeks after calving. Changes in condition can thus be used as a guide to the feeding of the cows.

Labour

No matter how well a herd is bred, or how scientific its feeding, it will not be a success unless there is a good person in charge of it. Good cowmanship is the foundation upon which successful dairy farming is built. A modern cowman has to be highly skilled, observant and conscientious. In addition, he must be prepared to milk twice a day, probably 6 days a week, at intervals as close to 12 hours as is practicable. The long hours and weekend work which this involves are much at variance with the working hours and conditions which are increasingly expected by workers in other industries. Fortunately, the job of working with cows can offer an interest and satisfaction which few others can provide. However, the fact remains that if a good man is to remain contented with such a demanding job—and if young people are to be attracted into it in future—it is essential not only to provide them with the basics of good wages and good housing, but also to give them the opportunity to do the job in a way which will satisfy their pride. Job satisfaction involves much more than just money and incentives, and goes far beyond even good working conditions; it demands leadership on the part of the employer, who must be prepared to let the staff share in decision making and understand the importance of what they are doing. Above all, the farmer must let the staff see that he is interested in every aspect of the herd.

Most dairy staff are paid a fixed wage which includes payment for regular overtime; in addition, they will normally receive free milk, some protective clothing and a house free of rent and rates.

Bonus and Incentive Schemes

In an effort to improve labour productivity and to give the staff a more direct financial interest in the success of the dairy enterprise, many different bonus and incentive schemes are operated. The most common scheme is related simply to milk production. The cowman may be paid either a flat bonus based on the amount of milk sold, in addition to the regular wage, or an increased amount per litre above a certain agreed level. Such simple milk bonuses have the disadvantage of encouraging the use of more concentrates, and to overcome this problem the bonus may be tied to some selected item, e.g. the margin over concentrates, in the monthly results of a costing scheme.

Other bonuses are aimed at tightening up the breeding programme. At its simplest, such a bonus may be a fixed amount paid to the first person on the farm to report a cow seen on heat; in its more elaborate forms the bonus may be tied to the annual calving index or may be a fixed amount for each cow which calves within, say, 390 days. Bonuses are also frequently paid for each calf reared.

To succeed, any bonus scheme must be simple both to operate and to understand, and must be paid at intervals which are regular and not too long. Most of the above schemes conform to those two requirements. However, there is a third and even more important requirement, which is that the incentive should be tied to some result which reflects the worker's efforts and is not affected by external influences. Clearly, none of the milk bonuses fulfils this requirement; animal disease, weather, quality of the forage, level of concentrate feeding and many other factors which are beyond the herdsman's control may have a much greater influence on milk production than anything which he can do. The breeding bonuses also fail to fulfil this requirement, though to a lesser degree. Thus bonus schemes can result in the herdsman receiving less money when the herd is going through a bad time for reasons over which he has no control—just the time, in fact, when he most needs encouragement.

There is a limited place for bonus and incentive schemes, but in general cowmen find motivation in the job itself provided that pay, housing and working conditions are satisfactory.

Relief and Contract Milking

Most dairy farmers can find somebody to milk the cows one day a week in order to give the cowman a break. The relief milker may be another stockman, a tractor driver, or the farmer himself. Holiday reliefs are usually provided in the same way, or with the help of a local relief milking contractor. Where emergency help is needed, for example because of illness or injury or where a cowman leaves at short notice, resort will usually be made to a relief milking service. These services will normally be able to send a relief man at short notice, but lodgings must be provided and travelling expenses paid and the service, while very useful, is expensive.

It was largely through such services as these that the contract milking system developed. A contract milker is self-employed, and instead of receiving a wage is paid an agreed price per litre of milk produced. The contract may be either between the farmer and the relief milking agency, which employs the milker on sub-contract, or direct between the farmer and the milker. The former arrangement is more expensive, but has the attraction that the agency is responsible for maintaining continuity of service in the event of illness or if the milker proves to be unsuitable. The terms of the contract may vary considerably, but in general the farmer undertakes to provide a minimum number of cows and to keep them provided with adequate supplies of food, litter and veterinary requisites. The farmer is also responsible for providing the milker with an adequate house, free of rent and rates. The contractor undertakes to look after the herd, to milk at stated times, to do basic veterinary work and to see that the animals are served at the correct time. Other work for which the milker is responsible will be detailed in the contract; this may include calf rearing, winter feeding, moving electric fencing, top-dressing pastures, and scraping slurry. The contractor normally finds and pays for his own reliefs.

Contract milking relieves the farmer of some of the managerial worries of a dairy unit, but tends to transfer some measure of control of the unit to the contractor. Even after taking into account the fact that he pays for his own insurance stamp and provides his own reliefs, a contract milker tends to cost considerably more than a cowman paid a normal wage.

Further Reading

Condition scoring of dairy cows, Advisory Leaflet 612, 1980, Ministry of Agriculture, Fisheries and Food, London

Report of the breeding and production organization, No. 32, 1981/82, Milk Marketing Board, Thames Ditton, Surrey

Wilson, B. and Macpherson, G., *Computers in farm management*, 1982, Northwood Books, Goswell Road, London

CHAPTER FIFTEEN

Breeding and Fertility

Inheritance—Systems of Breeding—Practical Application—Large-scale Breeding—Evaluating a Bull for AI—Choosing an AI Sire—The Oestrus Cycle—Heat Detection—Postpartum Interval—Conception Rate—Records—Control of Oestrus—Pedigree Cattle—Embryo Transplants—Culling

---

The object of a breeding policy, whether at national or at farm level, is to produce cows with a genetic potential for optimum milk production, efficiency of feed utilization, longevity and regular calving. All these and many other traits contribute to economy of production on the farm, and have their ultimate effect on the profitability of the dairy herd.

Cattle breeding is a very slow process; new genes cannot be created, and progress can only be made by the selection and recombination of existing genetic material. The immense difficulties of developing the ideal cow arise because there are so many different traits to be considered, and because many of the most desirable characteristics of cows are related inversely, e.g. milk yield and fat content. Milk production is affected by both breeding and feeding, and genetic differences are often concealed by variations in management. This was illustrated in a classic experiment conducted in New Zealand, when members of pairs of identical twin heifer calves were placed in herds with a record of either high or low yields. The performance of the individual twins followed the production of the herds in which they were kept, and it was concluded that the difference between herds was due mainly to management rather than to breeding. This does not imply that breeding is unimportant: as the level of herd management rises, the herd with the higher genetic potential should perform better.

Inheritance

Genetics may seem a complex and forbidding subject, but the basic principles are relatively simple. An animal's body is made up of millions of cells, which contain a fixed number of pairs of chromosomes; the exact number is specific to the species of animal. These chromosomes carry the genes, which are the submicroscopic bodies which govern the various characteristics of the animal; they are the units of inheritance. The sperm of the bull and the egg of the cow each contain a simple half of the genes of that animal. Thus each sperm or egg contains one of each pair of genes of the animal from which it comes, but which one is transmitted is a matter of pure chance and cannot be controlled. The calf receives 50 per cent of its genes from its sire and 50 per cent from its dam, and the calf's genetic make-up is a recombination of its parents' genes. Depending on which genes are transmitted, the calf may or may not show the same characteristics as its parents. Although like tends to breed like, it does not necessarily do so. The salient fact is that in order to judge how a bull's daughters will perform, reference must be made to the performance records of his earlier daughters. The only sure way of assessing the breeding merit of a bull is by a progeny test, and the only safe bull to use is a 'proven' sire which has undergone a successful progeny test.

Some characteristics, such as hair colour and the presence or absence of horns, are controlled by a simple pair of genes, but most traits of economic significance, such as milk production, milk composition, conformation and fertility, are governed by a large number of different genes. The genes are not affected by the environment in which the animal is kept, and 'acquired characters'—that is those characters which are imposed upon the animal, e.g. disease or injury—cannot be transmitted. Calves could be dehorned for scores of generations, but would never breed a polled animal.

Systems of Breeding

Inbreeding is the mating of animals which are more closely related than the average of the breed, examples being the mating of sire with daughter, dam with son, or brother with sister. Inbreeding tends to intensify characters, both good and bad, and hence it can be either valuable in the establishment of a pure line, where the parents are free from undesirable genes, or a disaster in producing deformed offspring and other undesirable characteristics. Inbreeding can be a valuable tool, but carries great risks if used indiscriminately. It does not, of course, create new characters any more than any other breeding system; it merely rearranges and intensifies those which are already there.

Linebreeding is a less extreme form of inbreeding, involving as a rule the use of a succession of related sires. Typically, son will follow father as the herd sire, and will thus be mated with his half-sisters. This system, while losing some of the effectiveness of inbreeding in fixing a desired characteristic, also avoids most of the risks of inbreeding. The system is widely practised.

Outcrossing is the mating of animals which are within the same breed, but which are not related for several generations. This system is frequently used to introduce some desirable characteristic in which the herd is deficient. The use of semen from the artificial insemination services usually involves outcrossing, unless the bull is related to the cows either by choice or by chance.

Crossbreeding is the mating of two different breeds, and is used where there are insufficient genetic resources within a breed to effect a desired improvement and to obtain the benefits of hybrid vigour. For some traits the offspring can be expected to be more or less intermediate in characteristics between sire and dam, but for other traits, such as vitality and fertility, the offspring may be superior to the average of the parent breeds and are said to exhibit hybrid vigour, or heterosis. It is often difficult to secure continuing benefits from hybrid vigour, but two breeding systems are commonly employed: firstly, backcrossing, which involves mating the crossbred cows to bulls of one of the foundation breeds, or secondly, crisscrossing, which involves the use of bulls of both of the foundation breeds and possibly of a third breed, the bulls of the different breeds being used in a regular sequence.

Practical Application

A major problem in breeding cattle is the slow turnover of generations. A period of at least 3·75 years elapses between a young bull's first service and completion of his first daughter's first lactation. Thus he must either be maintained, or a bank of his semen stored, until his breeding merit can be assessed. To increase the chances of finding a good sire, it would clearly be desirable to test more than one bull at a time, and the keeping of several bulls might well be required.

The second major problem of private breeders is the comparatively small number of animals in the average dairy herd. A progeny test conducted on 20 to 30 heifers in a single herd cannot compare in effectiveness with a test involving several hundred heifers in a large number of herds under differing systems of management.

A further problem is the limited scope for selection among the females in a herd. To maintain herd numbers, an annual replacement rate of 20 per cent should suffice; but 25 per cent is frequently required in practice. To procure this proportion of heifers, at least 50 per cent of the herd has to be bred pure, and to be safe it is usual to breed about 60 to 70 per cent pure, giving the option of selling a few surplus heifers. The remaining 30 to 40 per cent may be either bred pure or crossed with a beef bull. Table 15.1 shows that in a typical situation about 75 per cent of the animals culled from a herd are discarded for reasons other than breeding merit, which means that with a replacement rate of 20 to 25 per cent only 5 to 6 per cent of the herd are culled purely on breeding merit. Thus selection of females is not a matter of carefully choosing the best cows for breeding; two-thirds of a herd have to be bred pure simply in order to maintain herd numbers and the proportion rejected for breeding purposes is very small.

Table 15.1. Proportion of cows culled for different reasons

Main reason		Percentage of total number
Genetic		
Low milk yield and conformation		25·0
Disease		
Mastitis and udder damage	17·3	
Reproduction and repeat services	30·7	
Alimentary problems	2·7	
Metabolic diseases	4·1	
Locomotor troubles and accidents	5·1	
Other diseases	4·1	
	——	64·0
Age		11·0
		——
		100·0

(From Report No. 22, 1971–2, Milk Marketing Board, Thames Ditton, Surrey)

Large-scale Breeding

In cattle breeding, scale is immensely important, giving the numbers which permit selection from a large genetic pool, the computational facilities to process large numbers of records quickly and efficiently, and the resources to progeny-test numbers of young bulls. The principal cattle-breeding organizations in Britain are the Milk Marketing Boards, which are responsible for inseminating approximately 60 per cent of the total dairy cattle population. The Board in England and Wales inseminates in a year more than a million cows of the Friesian breed. Parts of the country are served by 'associated AI centres', which are mostly non-profit-making farmers' co-operatives offering a service which is similar to that of the Board but smaller in scale. Finally, there are the private breeding companies, which handle almost exclusively Friesian and Holstein semen. In these companies, several breeders normally combine to pool the genetic resources of their herds and to market semen from their own, and occasionally from other breeders', bulls.

Bulls for use by the Boards are purchased from private breeders largely through contract mating schemes. In these schemes, cows in private herds are selected for service by the best bulls in the stud of proven sires. These cows must be of good conformation and have above-average production records, and if a bull calf is born the Board has an option to purchase it.

Bulls for the artificial insemination service in England and Wales have to be approved by the Ministry of Agriculture, which only permits a limited number of inseminations for progeny-testing purposes. The bull is used as a yearling to get about 300 cows in calf in herds participating in a progeny-testing scheme. In this scheme, farmer members offer a specified number of cows for mating to young bulls and agree to rear and record the production of at least half the resultant heifers. In return, the members receive certain benefits. When the bull has completed his initial inseminations, he is not used again until the results of his progeny test are known.

Insemination of the cows is normally carried out by operators employed by the Boards or the associated centres, but it can also be done by farm staff in compliance with strict regulations relating to hygiene and the storage of semen. 'Do-it-yourself AI' is cheaper and gives a better choice of timing of service; these advantages have to offset the convenience of using the service provided by the AI centres, as well as the skill and experience of their full-time operators.

Evaluating a Bull for AI

To be accepted into the proven stud, a bull must satisfy a number of different standards, and five animals fail for every one which passes. A bull must be of good conformation and must have proved that he does not throw abnormal calves; his semen must be capable of extensive dilution and of recovering after being frozen, to give a satisfactory level of conception; and finally he is judged on the milk production and conformation of his daughters, as evaluated by the 'improved contemporary comparison' method. This technique compares the performance of a bull's daughters with the performance of the daughters of other bulls in the same herd at the same time. The genetic merit of the sires of the contemporary heifers is taken into account, and an allowance is made for the age of the heifers and their month of calving. The most reliable contemporary comparisons are those which are based on a large number of daughter records in many herds, and the number of these records is reflected in the figure of 'Weighting' which appears with the contemporary comparison figures. The higher the weighting figure, the more reliable is the test result. Progeny tests based on natural service cannot compare in reliability with those based on artificial insemination.

Choosing an AI Sire

Semen may be purchased from one of the private breeding companies, from one of the Milk Marketing Boards or from associated AI centres. A farmer using the AI centre of one of the Boards normally has the choice of three classes of bull:

The *bull of the day* is of the farmer's selected breed, but may be any of the bulls standing in the region. These are all proven sires, but are not of sufficient merit or popularity of prefix to qualify for the national list.

A *nominated sire* can be selected by name from a list of nationally available bulls, the cost of insemination rising in proportion to the demand for his semen. Supplies of semen from the more popular sires can be difficult to obtain and should be booked well in advance.

A *young bull* is the third alternative. This will be one of the bulls from the progeny testing scheme, which are not of course proven, so that their individual breeding merit is an unknown quantity and there is an element of gamble. However, these young bulls are of carefully selected breeding and past performance suggests that their average genetic merit is quite high. Semen from these bulls is cheaper than that from the other two classes of bull.

A producer may decide to use bulls of all three classes and balance the high cost of the most popular sires against the element of risk in using young unproven bulls.

In selecting a herd sire, the first priority will be a high contemporary comparison for milk, but consideration must also be given to the figures for fat and protein. To effect an improvement in milk yield the ICC of the sire should be at least twice the Cow Genetic Index of the dam. It is, however, a great mistake to choose a bull exclusively on figures. Good conformation affects such features as udder shape, teat shape and placement, ease of milking, and sound legs and feet, which have a direct influence on such important economic factors as the number of cows which one man can milk, and average herd life.

The various organizations which sell semen all supply full information on the merits of their bulls. The details given include the animal's pedigree, the price of an insemination and improved contemporary comparisons for milk, fat and protein (the two latter shown both in kg and as a percentage), with the weighting figures for each. There may also be an assessment of the conformation of the bull's daughters, made by a panel of breeders, and, where applicable, an indication that the bull carries a red recessive gene. This means that if he is bred with a cow which also carries the red recessive gene, the offspring has a 25 per cent chance of being red and white.

The Oestrus Cycle

Cows will only accept the bull during oestrus, i.e. heat periods, which normally occur at regular intervals of 17 to 24 days, the average interval being 20 days for heifers and 21 days for mature cows. The oestrus cycle (Fig. 15.1) is controlled by the pituitary gland, which secretes a hormone which stimulates the ovaries to produce eggs (ova). Every third week, the ovaries secrete a hormone (oestrogen) which induces oestrus. Near the end of this period, the pituitary gland produces another hormone (luteinizing hormone)

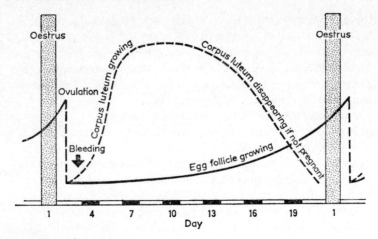

Fig. 15.1. Oestrus cycle in the cow
(After R. D. Frandson, *Anatomy and Physiology of Farm Animals*)

which stimulates the release of the eggs into the oviduct for fertilization. If conception takes place the corpus luteum does not disappear but continues to secrete the hormone progesterone, which stops the whole sexual cycle while the cow remains in calf.

Heat Detection

Spotting cows in season is essentially a matter of stockmanship; the best-managed cows in the world will never conceive at the right time if the staff do not notice them on heat and take the requisite action. The average length of a heat period for cows is 15 hours, and for heifers only 8 hours, but the length for cows can vary within a normal range of 12 to 28 hours. The old concept of the 'silent heat' is probably much rarer than is commonly supposed. A national average figure for apparent heat detection has been calculated at 60 per cent, but a realistic target is 80 per cent.

To reach this target, watching the cows only at milking and feeding times is not enough. The visibility of collecting yards from the parlour is often poor, and at feeding times cows are more interested in the feed than in each other. Separate observation periods are thus very important; they should each last at least 20 minutes and should include one at midday and one late at night, the latter being the best time of all.

When two cows are riding one another, it is the cow which stands still to be ridden which is on heat; this is the most positive way of identifying a cow in season. There are other signs of heat, such as dirty and steaming flanks, a swollen vulva with perhaps a slight discharge, and hair rubbed off the rump. Cows do not normally stand close together, thus any animals standing close, particularly if they are head-to-tail, should be looked at carefully.

Heat detection does not appear to be any more difficult in large herds than in small herds, since large herds are usually divided into groups either by milk yield or by calving date, and only one or two groups have to be observed at any one time.

Some herds have used vasectomized (i.e. sterilized) bulls to indicate cows on heat, but these bulls can transfer venereal disease and their use has gone out of favour. Of more general use are heat mount detectors; these consist of a clear plastic tube which is glued to the cow's tailhead, and turns a brilliant red when a cow has been ridden. A cheaper, and in some ways a better method, is to mark with coloured paste the tailheads of cows due to come into season.

When deciding whether or not to serve a cow, it is vital to know whether she has been served before, and if so, when. Cows sometimes show signs of heat when they are in calf, and inseminating a pregnant cow can upset the pregnancy. If in doubt, it is a good rule to serve the cow if it is a first service, but always to have a cow tested for pregnancy if her last service was 8 or more weeks before. For cows served less than 8 weeks previously, one can only rely on judgement and stockmanship.

Postpartum Interval

This interval is the number of days between calving and first service; there is a high positive correlation between the length of this interval and the calving index.

If the policy in a herd is to serve cows at the first heat after 56 days from calving, the average interval between calving and first service will be considerably longer—probably about 75 days. This is because the average first service would be at best half a heat period (i.e. 10 days) after the starting date and because all heat periods are not observed. Thus to achieve an average interval of 60 days between calving and first service, service should start at the first heat after 40 days from calving. The aim should be an average calving interval of 365 days and since the average gestation period for a dairy cow in calf to a dairy bull is 279 to 280 days, in an ideal situation every cow would conceive 85 days after calving. This is clearly impossible, but to allow for missed heat periods and normal conception rates, it is advisable to aim to serve every cow at between 50 and 60 days after calving. It is recommended that service should commence 40 days after calving, but not earlier. The reproductive system of the cow must be given time to settle down after calving, and conception rate drops sharply for services at less than 40 days. Delaying first service until the first heat after 60 days will result in a slightly higher conception rate but a longer calving interval. The dry period before calving should be about 56 days for cows and 70 days for heifers, but, as with the postpartum interval, this depends on the condition of the animal.

Conception Rate

The average calving interval of cows in Britain is about 395 days, and has not altered significantly in the last 30 years. Reasons advanced for this lack of improvement include increasing herd size, reduced use of labour per cow, and changes in housing systems. In herds with an unsatisfactory calving interval, a poor conception rate is often cited as the main reason, but the trouble is more likely to lie with poor heat detection and an unduly long postpartum interval. Conception rate is nonetheless of great importance and, to a considerable extent, within the control of the farmer. For reasons which are still largely undefined, prenatal mortality, i.e. death of the foetus between conception and birth, is high and has been estimated to be between 20 and 50 per cent. Thus, the proportion of services which will ultimately result in a birth cannot exceed 80 per cent, which sets a ceiling on what is physiologically possible.

There are various methods of measuring fertility. Non-return rate at 30 to 60 days measures the percentage of first services which have not been repeated at between 30 and 60 days. This method is quick and simple, but it takes no account of either prenatal mortality or poor heat detection after 60 days, and a true conception rate (i.e. percentage of cows which calve to the first service) is probably at least 20 per cent lower. Thus the average non-return rate of the artificial insemination service of 80 per cent probably represents a true conception rate of 53 to 55 per cent.

It is difficult to make a fair comparison between results from artificial insemination and natural service, since natural services are not always recorded, but it seems probable that the difference, while showing a small advantage to natural service, would not be significant compared with the influence of nutrition and other managerial factors.

Conception rate can be reduced by diseases such as brucellosis, trichomoniasis and vibrio foetus, but these diseases are of declining importance, and more common troubles are infections of the uterus following difficult calvings or retained cleansings, and hormonal disturbances. Conception rate will also be affected by the length of the postpartum interval, but by far the most important factor influencing conception rate is nutrition. One of the tenets of traditional farming wisdom was that animals conceive best on a rising plane of nutrition—an example is the old practice of 'flushing' ewes for tupping—and there is much evidence to show that cows have a higher conception rate when they are gaining weight rather than losing it. This fact poses a considerable problem with a high-yielding cow, which will be served at the peak of lactation when her appetite will be only slowly recovering from the post-calving depression. At this time, the limited amount of food which she is able to consume has to support a high milk yield, to replace some of the bodyweight lost at calving, and to enable her to conceive. Her ration must have a high ME and the correct content of protein and minerals. The calcium: phosphorus ratio is particularly important. Many foods, e.g. kale, lucerne and sugar-beet pulp, are high in calcium but low in phosphorus, and a phosphorus deficiency is a common cause of low conception rates. Mineral mixtures high in phosphorus are important to balance calcium-rich foods. Copper deficiency is a problem in some areas.

Conception rate may be reduced by stress such as extremes of weather, a change of herd companions or of milker. There is no evidence that conception rate suffers as milk yields rise until extremely high levels are reached.

The timing of service is important, but the difficulty is to know exactly when oestrus commenced. Ovulation takes place 10 to 15 hours after the end of oestrus (Fig. 15.1), and this is the optimum time for service. In practical terms, the best advice is to serve cows towards the end of the period of standing heat, bearing in mind that this period will be shorter during the winter months.

Infertility should be regarded as a herd problem rather than an individual problem, and a herd should be considered to have a fertility problem if it has a conception rate of less than 50 per cent, a calving index of more than 385 days, or if any cow needs four or more services. Such a herd should take veterinary advice, though the remedy is more likely to lie in the fields of nutrition and general management.

Targets to aim at are a calving index of 365 days, with a culling rate for reproductive causes of 4 per cent, and 96 calvings per 100 cows per year.

Records

There is no aspect of herd management in which records are more essential than the breeding programme. The keeping of records was covered in Chapter Fourteen, and now their use will be considered. The importance of recording all observed heats has been stressed; if a note is made of the date of the first heat of a freshly calved cow, a particularly careful watch can be kept on her about 3 weeks later and the chances of a heat being missed are greatly reduced.

Visual aids, or cow calendars, are particularly useful in pointing out cows which have not been seen on heat 40 days after calving; all such cows should be seen by the veterinary surgeon as a routine at his next visit. Records will also pinpoint cows which are cycling irregularly, i.e. coming on heat at other than approximately 3-weekly intervals; again such cows should be examined by the veterinary surgeon.

In many herds, a routine pregnancy diagnosis is made of all cows about 8 weeks after service. As an alternative, use may be made of the Milk Marketing Board's postal pregnancy diagnosis service, which merely requires a milk sample taken 24 days after insemination. Pregnancy diagnosis is by measuring the level of progesterone in the milk. An accuracy approaching 100 per cent is claimed in non-pregnant cows, and about 90 per cent in pregnant cows.

Control of Oestrus

Oestrus can be controlled and synchronized by the use of the hormone prostaglandin and its synthetic analogues. When these are injected into the cow it causes the corpus luteum to atrophy and induces heat. It will not affect cows at all stages of the breeding cycle, and the usual practice is to give two injections at an interval of 11 days. This ensures that the second injection will bring all normally-cycling animals into oestrus. The insemination is usually done twice: first 3 days after the second injection, and the second insemination on the following day. An alternative is the use of PRID (Progesterone-releasing intravaginal device) coils.

Such techniques are particularly useful where it is preferred to inseminate a group of heifers rather than use a bull. All the heifers will come on heat and be inseminated on the same day, and although they will not all conceive to the first service, a tight calving pattern can be achieved without the problems of heat detection which are normal with heifers.

With dairy cows, the technique may be used strategically on individual animals to make up for deficiencies in heat spotting. However, this technique is not a cure for infertility and it is not cheap. Controlled ovulation will not bring a cow on heat which is not already cycling normally, but it will bring a cow into season at a selected time, which is a great help in overcoming the problem of the so-called 'silent heat'. A typical use is on the later-calving cows in a herd with a close calving pattern, e.g. a spring-calving herd.

Pedigree Cattle

Pedigree means known and recorded ancestry, and pedigree cows are animals which are bred from pedigree parents and registered with the breed society, which keeps a full record of the parentage of all registered cattle of that breed. Entry to the register can be gained by 'grading up' animals which meet certain standards of appearance and are sired by a pedigree bull and, in some breeds, have satisfied certain standards of milk yield and fat. In three generations, the offspring of these grading-register cattle can become full pedigree.

Embryo Transplants

It is possible, by injection, to induce a cow to produce a large number of eggs (6 to 16) at a single ovulation, which is known as 'super-ovulation'. By careful artificial insemination these eggs can be fertilized and then subsequently collected without a surgical operation. After inspection under a microscope, the eggs, with a minor surgical operation, are implanted into recipient animals, whose oestrus cycle has been carefully controlled to be at the optimum stage for acceptance. Alternatively it is possible to freeze the eggs and store them before use. The average rate of successful transplant of such embryos is about 6 per ovulation.

The attraction of this embryo transplant technique is that it permits cows of superior genetic merit to produce far larger numbers of offspring than would be possible by natural means. It can also ensure the production of at least one male offspring in a contract mating for bull breeding. However, the high cost seems likely to limit the impact of this technique.

Culling

The culling rate in the national dairy herd is between 20 and 25 per cent, and the average cow has a life in the herd of only three to four lactations. Ideally cows would only be replaced either because of old age or because they were of below-average genetic merit, but in fact many other reasons arise for their disposal.

Reasons for culling in 14 herds with a total of 1,350 cows were analysed and are shown in Table 15.1. The most noteworthy feature of the table is the low proportion—25 per cent—of the total number of cows culled which was voluntary, i.e. for genetic reasons, whereas 64 per cent were unplanned cullings. Reducing the proportion of involuntary cullings is an important aim of management. The high proportion culled for reproductive reasons included calving troubles, abortions and infertility. Locomotor troubles include bad feet, arthritis and other troubles of the legs and feet.

Further Reading

An introduction to cattle breeding, Booklets 2403, 2404 and 2405, 1982, Ministry of Agriculture, Fisheries and Food, London

Bowman, J. C., *An introduction to animal breeding*, Studies in Biology No. 46, 1974, Edward Arnold, London

Brackett, B. G., Seidle, G. E., and Seidle, S. M., *New technologies in animal breeding*, 1981, Academic Press Inc. (London) Ltd, Oval Road London

Dalton, D. C., *An introduction to practical animal breeding*, 1980, Granada Publishing, London

Esslemont, R. J., 'Management with special reference to fertility', Chapter 13 in *Feeding strategy for the high yielding dairy cow*, (eds. Broster, W. H. and Swan, H.), 1979, Granada Publishing.

Frandson, R. D., *Anatomy and physiology of farm animals*, 1975, Lea and Febiger, Philadelphia

Hunter, R. H. F., *Reproduction of farm animals*, 1982, Longman Group Ltd, Harlow, Essex

Johansson, I., *Genetic aspects of dairy cattle breeding*, 1961, University of Illinois Press, Urbana

Swan, H. and Broster, W. H. (eds.), *Principles of cattle production*, 1976, Butterworths, London

CHAPTER SIXTEEN

Herd Replacements

Replacement Policy—Age at First Calving—Calving—Calf Feeding—
Calf Management—Rearing to 1 Year—Rearing from 1 to 2 Years—
Spring-born Calves—Target Liveweights—Calf Health—Salmonella
and Virus Pneumonia—'Worms'—Rearing and Profits

The average rate of replacement in dairy herds is 21 heifers per 100 cows each year. If the herd size is to remain static, and undue culling is not to take place, then a minimum of 60 per cent of the herd must be mated to a dairy sire (Chapter Fifteen). Replacement rates vary between herds and between years depending on the incidence of disease, the price of cull stock and whether the herd is increasing in numbers. The age at first calving also has a marked effect on the number of replacements on the farm at any one time. For example, with a 20 per cent replacement rate there will be 40 replacements on a farm with a herd of 100 cows if the age of first calving is 2 years, but 60 replacements if calving is at 3 years.

An indication of the annual rate of replacement per 100 cows for herd lives of different lengths and three levels of cow mortality is given in Fig. 16.1. A high rate of replacement is not necessarily an indication of poor herd management, as the farm policy may be to have only young and fit animals in the milking herd and to sell the older cows. In planning a replacement policy it must be remembered that for every 100 cows mated, there is an average production of only 83 live calves per year owing to conception failures, abortions, stillbirths and difficult calvings.

The selection of heifer calves as herd replacements has been discussed in Chapter Fifteen, and it is stressed that the performance of the dam, sisters and half-sisters should all be considered when keeping a calf.

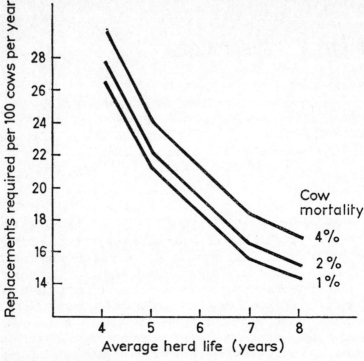

Fig. 16.1. Number of replacements per 100 cows, depending on herd life and cow mortality
(From *Rearing replacements for beef and dairy herds*)

Replacement Policy

Replacements for the dairy herd can be home-reared, contract-reared, or purchased. Home rearing makes it possible to breed for improved yield and conformation, and there is no risk of introducing disease on to the farm, but land and resources are needed for this extra enterprise. Labour and management must be devoted to rearing replacements, and there is a considerable capital requirement for the loss of calf income and rearing costs. Land suitable for milk production may have to be diverted for use by replacements, and an overall loss of profit may result.

Contract rearing is not widespread, and the system requires a precise agreement between the owner of the dairy herd and the rearer. The herd owner can still select his calves, but land and labour are not spent on rearing replacements. The rearer will be a specialist who can devote time and skill to this task, and there is not a high risk of disease being transmitted to the original herd. However, contract rearing requires considerable capital, and it is not easy to find a competent rearer.

If replacements are purchased, either privately or in a market, there is always a danger of introducing disease into the herd, but no labour, management time, and land have been spent on rearing. Capital has not been invested in non-productive stock; it may be possible to expand other, more profitable, enterprises; and the animals can be selected and rejected at the time of calving. To offset these advantages are the facts that the cost of purchases is often unpredictable, and that no breeding policy can be followed. The selection of the correct replacement policy for a specific farm will depend on individual circumstances, and must be ultimately measured within a whole farm situation.

Age at First Calving

Early calving not only requires fewer replacements on the farm, but generally results in a higher lifetime production of milk from the cow. In the past, there was much confusion on the subject of early calving because of the interaction of age and weight. Many animals which calved early were also low in weight, but it is now clear that if the animal is well grown, early calving is not a disadvantage. As shown in Table 16.1, if factors such as the number of lactations and lifetime yield are considered, any delay in calving after about 24 months should be avoided. Yield in the first lactation is invariably reduced by earlier calving, but lifetime yield is increased.

Table 16.1. Age at first calving and subsequent milk yield

	Age at first calving (months)				
	23–5	26–8	29–31	32–4	35–7
Herd life (years)	4·0	4·0	3·8	3·8	3·8
Milk yield					
Lifetime (100 kg)	187	187	180	180	176
Per day in herd (kg)	13·1	13·2	13·1	13·1	13·2
Per day of life (kg)	8·8	8·4	7·8	7·5	7·3

(From *Rearing replacements for beef and dairy herds*)

The incidence of difficult calvings in heifers is usually double that in mature cows, and is increased when heifers are calved at under 24 months old. Calving difficulties are affected by the breed of bull, and even by the individual bull within a breed. Large beef breeds such as the Charolais, Simmental and Limousin cause much higher levels of calving difficulty when mated to Friesian heifers than the smaller breeds such as the Aberdeen Angus and Hereford. These two latter beef breeds will also colour-mark the calves. Friesian bulls give more calving problems than Hereford and Aberdeen Angus bulls, and should only be used to mate young heifers in special circumstances unless the bull is known to breed small calves. The heifer itself can influence calving difficulties if it is bred from a line containing large sires. Thus, although calving at 24 months has many economic advantages, it is important to have the heifer at the correct weight and condition and to use the appropriate bull.

Calving

Successful calf rearing starts with the management of the cow. If the date of service has been recorded, the date of calving can be calculated from the average gestation length of 279 to 280 days, and appropriate plans made for calving. Outdoor calving in a small sheltered field without ditches and other hazards is ideal in summer, but loose-boxes (Chapter Twelve) are preferable at other times. Depending on the calving pattern, one box is required for approximately every 20 cows in the herd, and these should be situated where the animals can be easily and regularly inspected. The boxes should be clean and fairly well bedded with straw. Calving in a byre or in a cubicle house is not recommended, as both the cow and calf can be subject to injury and can disturb the rest of the herd.

The cow should be put in the box 2 to 3 days before calving, when the teats and udder will have distended noticeably. About 1 day before calving, the ligaments on the sides of the tailhead loosen and the vulva becomes enlarged and flabby. Food intake falls at this time, but water should always be available. At the time of calving, the water bladder appears first and then, in a normal calving, the forefeet with the muzzle of the calf resting on them. If assistance is necessary, clean calving ropes attached to the legs of the calf are helpful, and the pulling must be in rhythm with the contractions of the cow. If real difficulties arise, either because of abnormal presentation of the calf or lack of room for it to pass through, skilled veterinary assistance must be obtained without delay. Immediately after birth, the foetal membranes and the mucus from the nose and mouth should be removed, and artificial respiration applied if the calf is not breathing. The cow will normally start to lick its calf at once.

Every effort should be made to see that the new-born calf receives colostrum, i.e. the milk secreted in the first few days after calving. Colostrum is a mixture of true milk and certain constituents of blood plasma which have been concentrated ten- to fifteen-fold in the udder before calving. It has a high content of protein, particularly the immune lactoglobulins and antibodies which protect the calf against disease in early life, and of carotene and vitamins A, D and E. The blood of the newly born calf contains no antibodies, and it is imperative that the calf receives colostrum in the first 6 to 12 hours, as after this time the antibodies cannot pass into the bloodstream. The colostrum and the antibodies are more efficiently absorbed by suckling than by bucket feeding. Thus, calves may either suckle their dams for 2 days after birth or, if calves are removed from their dams at birth, the appropriate colostrum should be given to the calf. Colostrum stored in churns will acidify rapidly and can be used for calf feeding, but not in the first few days of the calf's life.

If a calf is deprived of colostrum, and manages to survive, it will develop its own immunity after about 10 days of age. Colostrum from another cow in the same herd can be valuable for a calf unable to have its own dam's colostrum, and the following recipe can be given as a last resort. Whip up a fresh egg in 0·6 l whole milk plus 0·3 l warm water, and add 1·5 ml castor oil. This mixture should be given three times per day.

Calf Feeding

There are numerous systems of calf feeding, ranging from ones using large amounts of liquid food to others with the minimum of liquid so that there is an early development of the calf's rumen. Systems containing liquids enable diets of high digestibility to be given, with subsequent large weight gains, and are used for veal production. Systems of dry feeding are preferable for rearing normal herd replacements, although the foods used are lower in digestibility than liquid diets.

Early weaning at 5 weeks of age is a highly satisfactory system of dry feeding, and has the combined advantages of a low labour requirement and fewer risks to the calf from infections and metabolic diseases than liquid feeding. Briefly, the early weaning system consists of feeding a limited amount of either milk or milk substitute for 5 weeks after birth, and encouraging the calf to eat concentrates and hay. The amount of milk substitute is restricted to a maximum of 3·5 l per day, which is about sufficient for the maintenance requirements of the calf.

The concentrate on offer to the calf should contain 17 to 18 per cent crude protein and be of high palatability, and the animals will eat gradually increasing amounts as they become older. The concentrates can be home-mixed, but it is generally preferable to purchase this specialized food. The early-weaning concentrates are normally supplied as a cube and should be given *ad libitum* in a clean, dry container which is emptied and refilled every few days. An early interest in the cubes can be stimulated by putting them in the bottom of the pail which has contained the liquid feed. The hay on offer must be free of dust and moulds, and of the highest D-value available.

Weaning from the liquid feed should be abrupt unless the calf is poor and weak through scouring. At this time a healthy 5-week-old calf will be eating about 0·7 kg concentrates per day. Feeding the milk and milk equivalent three times per day has no major advantages over two feeds per day, and it is now a widespread practice to give only one liquid feed per day. The system of once-per-day feeding is economical on labour, and highly successful if a high-fat milk equivalent is mixed with warm water in the ratio of 1 : 7 or 8 so that each calf receives the equivalent of about 0·4 kg of dry milk equivalent per day. The milk equivalent powder should contain 20 per cent oil, 25 to 26 per cent protein and no fibre, and should mix readily with the hot water.

The single feed can be given at the most convenient time of the day, but should not vary from one day to another. The proportion of water in the mixture can be varied to ensure that the calves consume the correct amount of the high-fat milk equivalent. In any early-weaning system, the calves should have access to fresh, clean water from birth; this is vital if liquid feeding is restricted and concentrate intake is to be encouraged.

If calves are kept in separate pens, pail feeding is a quick and easy system of feeding the milk equivalent. Other advantages of single pens include a lack of competition for the liquid feed and a reduced chance for the spread of disease. However, if large groups of calves are penned together, an automatic feeder can save much time. These machines vary in size and complexity, but it is essential to keep them well adjusted, clean and thoroughly disinfected.

Calf Management

Calves should, if possible, be reared in batches, so that labour can be concentrated on one enterprise at one time, and also to ensure that the calf house is empty for periods of at least 1 month between batches. This allows the building to be cleaned and disinfected, and will assist in breaking any build-up of disease, even in a sub-clinical form. Steam-cleaning is particularly effective for calf pens, wooden slats, and the entire contents of the calf house.

Dehorning should be done before the calf is 3 weeks old. Caustic potash in the form of a stick, and collodion solutions, will remove the horn buds if used carefully, but there are fewer problems with cauterizers heated by either gas or electricity. The electric cauterizer must be properly earthed, and, when warmed up, should be held around the horn bud for 5 to 10 seconds, so that the tissue is seared to a depth of about 2 mm. An anaesthetic must be used for disbudding calves unless the operation is done by chemical cautery in the first week of life.

Extra teats on an udder can be a nuisance, and spoil its appearance. Such teats should be removed carefully before the calf is 3 to 4 weeks old by cutting with a pair of sharp, curved scissors which have been sterilized by boiling.

Calf identification is vital from birth onwards (Chapter Fourteen), and this must be done to fulfil the requirements of the breed society and other authorities. Ears can be tattooed and clipped to form a number code, but the easiest and clearest system is the use of eartags. These are made of plastic, are easy to insert in the ear, and are simple to read. By the use of coloured tags it is possible to identify different age groups, and even to differentiate between various herd sires. Eartags are lost occasionally, and new ones should be inserted at once to ensure that there is no problem with identification. Coloured and numbered tags are preferable to either neck bands or freeze-branding, which can be used with older animals.

When the calf is young, any abnormalities should be noted, and the calf sold at once for slaughter if it is not up to the standard of the herd. Small umbilical hernias are fairly common and should not be a reason for rejection.

Rearing to 1 Year

If an autumn-born calf is to calve at 2 years, there must be an average liveweight gain of 0·6 to 0·7 kg per day throughout the rearing period. In the first 5 weeks of life the gain should average about 0·4 kg per day, and increase to 0·6 kg per day from weeks 5 to 12. In this period, the concentrates should be given *ad libitum* up to a maximum of 2 to 3 kg per calf daily, depending on the target liveweight gain. Concentrates containing 13 to 14 per cent crude protein, and hay of high D-value, must be available. At weeks 8–10, silage of high D-value can be given in small amounts, so that the animals become familiar with this important food.

After week 12, calves can be given 1·5 to 2·0 kg per day of a concentrate containing 12 per cent crude protein if hay of high quality is available. Silage, which was given in amounts not exceeding 1 to 2 kg per day at 12 weeks old, can now be increased slowly, but it is essential to feed only material of the highest D-value and without any moulds. From birth to turnout in the spring, the average liveweight gain should be 0·7 kg per day.

In the first summer's grazing this same rate of gain should be achieved by feeding an unbroken sequence of highly digestible leafy grass, and by having an adequate control of parasitic worms. Calves tend to graze selectively, and if they are compelled to eat a large proportion of the herbage, growth rates will be poor. A calf-grazing system should therefore involve some form of rotation with a high degree of selection.

An excellent method is to graze calves in their first summer immediately ahead of a group of heifers in their second grazing season. The young calves can graze in a highly selective way and consume the grass with the highest nutritive value, and then the older heifers will eat the residual herbage of lower feed value. Stocking rates should range from 3·5 calves plus 3·5 heifers per ha where no conservation crop is made, to 2·5 calves plus 2·5 heifers per ha where some of the area is conserved. The system should have 8 to 12 separate paddocks with a rotational grazing system of about 4 weeks.

In practice, this system can utilize over 90 per cent of the available herbage, control the worm burden of the calves, and give liveweight gains of 0·8 kg per day. Good fencing is essential, and the cattle should be moved to the next paddock according to the requirements of the older heifers. This ensures that the younger animals are not short of herbage. The daily feeding of about 1 kg concentrates containing 12 per cent crude protein is generally worthwhile for about 2 weeks in spring, and again in autumn if grass is scarce. At no time should the growth rate decline if the animals are to calve at the correct weight at 2 years of age.

Rearing from 1 to 2 Years

The second winter can be considered in two parts. In the first part, the housed heifers should maintain a liveweight gain of 0·6 kg per day to ensure a target weight of about 330 kg at the time of mating. Silage and hay of good quality, supplemented with 2 kg per day of either barley or concentrates containing 12 per cent crude protein, should form the basis of the ration. If silage with a D-value of 68 to 70 is offered, concentrates can be virtually eliminated. After the heifers are in calf, the liveweight gain can be reduced to around 0·5 kg per day with an increased reliance on forages of high quality and a reduced use of concentrates. Other foods such as roots, beet pulp, and maize silage can also be given at this period.

During the second grazing season, the target weight gain should again be 0·7 kg per day, and this can be achieved without any concentrate feeding if the standard of grassland management is high. The animals should never be short of highly digestible leafy herbage, which can be achieved by rotational grazing and the use of fertilizer nitrogen. At the end of the 2-year period of rearing a Friesian heifer should weigh about 500 kg and be in a fit condition for calving.

Spring-born Calves

Calves born in the spring cannot make as efficient use of the first-summer grazing as calves born in the autumn. Because of this difficulty, the early ration of the spring-born calf has to contain a high proportion of concentrates plus forages of the highest D-value if calving at 2 years of age is to be achieved. The rearing of the spring-born calf is identical to that of the autumn-born calf for the first 3 to 4 months. Outdoor rearing of young calves at grass is not a popular system, especially if the calf house is empty. Young calves aged 3 weeks can digest grass as efficiently as adult animals, but the animals rarely eat sufficient grass to achieve the desired liveweight gains. The standard of grassland management has to be high, and some form of temporary housing is desirable. Labour demand is high, with concentrates and milk substitute having to be transported to the field, and on balance it is preferable to house and feed the calves in the normal way.

Many calves are not grazed at all in their first summer, but if the weather is suitable and grass is available, calves can be put out to grass at 12 to 16 weeks of age. Supplementary feeding will be required unless the calves have access to ample leafy grass of high D-value, if a growth rate of 0·6 kg per day is to be achieved. A similar rate of gain is necessary throughout the first winter, and 0·7 kg per day during the following summer, if the target weight is to be reached at service at 15 months. At the end of the second grazing season, the liveweight gain can be allowed to decline to about 0·6 kg per day, and this same rate should be maintained during the second winter. The forages offered at this time must be of high quality if undue reliance is not to be placed on concentrates, which should not exceed about 2 kg per animal per day.

Target Liveweights

Target liveweights for various breeds are given in Table 16.2. It is important to achieve the correct weight at the time of service, and also the correct rate of gain, as this also affects the rate of conception.

Table 16.2. Target liveweights at service and calving

| | Weight (kg) | |
Breed	Service	Pre-calving
Jersey	230	335
Guernsey	260	390
Ayrshire	280	420
Friesian	325	510

(From *Rearing replacements for beef and dairy herds*)

If calving at 2·5 years is preferred to calving at 2 years, then all the growth rates given earlier can be reduced by about 0·1 kg per day, but with target weights which should be 25 to 50 kg higher than those in Table 16.2. It should be possible to feed a smaller weight of concentrates when calving at 2·5 years compared with 2 years, but forage quality and grazing management must still be high.

Calf Health

On average, of every 100 calves born alive, about 6 die before they are 6 months old, and three-quarters of the losses occur during the first month of life. In addition to these deaths, there are considerable economic losses as a result of ill-health in calves which do not die. Mortality rates are highest in February, March and April, after a steady increase in the rate from September onwards. This suggests that there is a gradual build-up of disease in the calf house, and emphasizes the need to have the building completely empty between different batches of animals.

Many of the health problems in young calves are associated with the bacteria *Escherichia coli*, and certain strains may become dominant so that the colostrum is not so effective as normally. Nonetheless it is vital that the calf receives colostrum so that it develops a passive immunity. The level of immunity may be inadequate to prevent the multiplication of the *E. coli* in the intestine of the calf, and it is then important that the diet should be correct. Any reduction in the efficiency of digestion will allow incorrect intestinal flora to become established. For this reason, milk for calf feeding should be only gently heated, so that the casein forms a clot in the abomasum and thus allows the whey and the digested casein to pass slowly into the small intestine. If either milk or milk substitute is overheated, little clotting occurs in the abomasum, so that the undigested casein passes to the small intestine, providing conditions for the multiplication of *E. coli*, and scouring occurs. This condition decreases the liveweight gain and, if severe, will cause death.

To reduce the risk of scouring it is important to feed only fresh milk at body temperature (36 to 38 °C), and a purchased milk substitute from a reputable source. Overfeeding and sudden changes in diet must be avoided in calves under 3 weeks old, and all pails and equipment used for feeding must be cleaned and sterilized. If a calf is scouring, it is helpful to starve the animal for one feed and then offer only a small amount of milk to maintain the liveweight. This low level of feeding should be continued until the faeces of the calf become firmer, when more milk can be offered. Scouring calves lose excessive amounts of water, and there is a risk of dehydration which must be rectified as rapidly as possible.

Salmonella and Virus Pneumonia

Calves can also be infected by the *Salmonella* organisms, which produce a scour which is often bloodstained. Severely affected calves become emaciated and weak, and eventually die. To combat this disease, the source of infection must be detected and removed and the healthy calves vaccinated 1 week before exposure to the disease. An adult cow can be a carrier of the disease, and there is a risk that slurry can infect pasture. Purchased animals should be segregated from other animals on the farm, and all slurry should be stored for at least 1 month before applying it to pasture.

In recent years there has been an increase in the incidence of 'virus pneumonia' in housed calves. This problem usually occurs in the winter months, with symptoms which can include a nasal discharge, a dry cough and respiratory distress. Coughing after the calves have exerted themselves is often the major symptom. Poor ventilation and high humidity in the calf house are widely blamed as predisposing causes, and evenly distributed inlet ventilation plus outlets at a high point should provide the gentle air movement which is so essential (Chapter Twelve). Treatment should be aimed at reducing any secondary infection, plus correct feeding and the avoidance of draughts and wet bedding.

'Worms'

Husk, hoose and lungworm are various names for a condition caused by a heavy infestation of the lungs with a nematode worm (*Dictyocaulus viviparus*). This condition occurs when the animals are grazing, and the main symptom is a cough which varies in severity according to the degree of infection. There is a loss of appetite, a staring coat and even death in severe outbreaks. The disease is now controlled by an oral vaccine ('Dictol'), which is given to the calves in two separate doses before they are turned out to pasture. However, even with vaccinated calves it is worthwhile to keep the animals well fed and to avoid grazing areas which are heavily infested with worms.

In some districts it is better to keep the calves indoors during their first summer, but a small exposure to the disease will help to build up an immunity.

Parasitic gastro-enteritis is another disease caused by parasitic worms, mainly *Ostertagia* spp., which infect the abomasum and small intestine. The symptoms are a progressive loss of condition and scouring, and in severe infestations deaths may occur. Calves are highly susceptible, but older cattle can also become infected. Although the disease is associated primarily with grazing, poor growth rates can be recorded in the following winter, since the digestion of the infected animal can be impaired. Cattle which have never been at grass do not usually have a worm burden, but once young cattle are grazing, the number of worms can increase rapidly. The disease can be prevented by adequate feeding and a combination of good grazing management and dosing. If the animals are set-stocked and poorly managed, dosings will be required about 1 month after grazing starts, again in July and finally in autumn before housing. Alternatively, dosing can be done in July or August to coincide with a change to a fresh, uninfested pasture such as a silage aftermath.

If calves are always grazed ahead of the older replacements, there may be little need for dosing, especially if a clean young ley is being grazed. Any loss of condition due to worms indicates a failure in management which should be remedied at once. When animals are dosed, all animals in a group must be treated, and the group moved to a clean pasture. In general, calves should not be exposed to a sudden heavy infestation of worms, and should be managed so that they ingest a gradually increasing amount of high-quality grass.

Rearing and Profits

Calf health is often a neglected aspect of dairy farming because of an undue emphasis on the adult animals which are the immediate source of income. In the long run, the supply of healthy, well grown replacements is paramount for the future success of the dairy enterprise, and every attention should be paid to the health and well-being of the calves and young growing stock on the farm. Veterinary advice should be sought quickly for any disease problems, but good housing, correct feeding and careful attention are essential for successful calf and replacement rearing.

Finally, if the efficiency of rearing replacements is increased by lowering the age of first calving, increasing the rate of stocking, and reducing losses, more dairy cows can be kept on the farm. This can dramatically increase total farm profits, as land is more profitably utilized by dairy cows than by .herd replacements. The rearing of heifer replacements is an aspect of dairy farm management which will handsomely repay the investment of time, attention to detail and the application of basic knowledge.

Further Reading

Kilkenny, J. B. and Herbert, W. A., *Rearing replacements for beef and dairy herds*, 1976, Milk Marketing Board, Thames Ditton, Surrey

Mitchell, C. D., *Calf housing handbook*, 1976, Farm Buildings Information Centre, N.A.C., Stoneleigh, Warwickshire

Roy, J. H. B., *The calf*, 4th ed., 1980, Butterworths, London

CHAPTER SEVENTEEN

Cow Behaviour and Health

Grazing and Eating—Ruminating and Lying Down—Walking and Other Activities—Social Rank—Normal Health—Dung—Other Indicators—Milk Yield—Preventive Medicine—Mastitis—Cell Counts—Control Hygiene—Summer Mastitis—Milk Fever—Grass Staggers—Acetonaemia—Bloat—Lameness—Metabolic Profile Test

A knowledge of the normal behaviour and the daily activities of dairy cows can be useful in two main ways. First, a change in behaviour can indicate some fault in management or animal health, and second it may be possible to alter and improve the herd routine as a result of studying the pattern of behaviour. Part of the stockman's art is a careful observation of the behaviour of his animals so that any abnormalities can be detected, and remedial action taken swiftly.

Dairy cows have a fairly regular pattern of behaviour from day to day, the main activities being eating (including grazing), lying down, idling, and ruminating (Fig. 17.1). Cows rarely sleep, i.e. completely lose consciousness. An animal may lie with its eyes closed, but noise and movement will cause an immediate response. It is claimed that the thorax must always be kept in a vertical position to enable the rumen to function properly, and hence the cow rarely rests flat on its side.

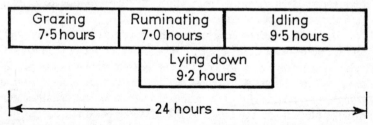

Fig. 17.1. Length of time of the main activities of a grazing dairy cow (Based on R. Waite, in *Animal Health, Production and Pasture*)

Grazing and Eating

On average a cow spends about 6 to 9 hours per 24 hours grazing at pasture (Fig. 17.1). This time includes the entire period spent searching for the grass, biting the herbage and swallowing. Under normal climatic conditions, grazing time is about equal in each of the two periods between milkings, and it is therefore important to offer herbage of similar quality and quantity for both day and night grazing. During periods of abnormally hot and sunny weather, grazing may be restricted during the day, with most of the grazing between the evening and morning milkings.

Some pastures offered to cows at night are bare and overgrazed, which can restrict the animals' intake of herbage. In both the morning and the evening, the main period of grazing is immediately after the cows have entered the pasture, and most grazing stops about 1 hour after sunset. As the amount of herbage available to the cow declines, grazing time increases. In one typical observation, grazing occupied 7·7 hours per day when cows first entered a fresh paddock, 9·4 hours on the second day and 9·8 hours per day on the third and fourth days. The quality of the herbage does not have such a marked effect on grazing time, but it affects the ruminating time, and hence the ratio between grazing and ruminating times. Stemmy herbage with a high content of crude fibre causes a long ruminating time, whereas leafy material with a low content of fibre is more easily broken down and ruminating times are shorter.

Feeding concentrates to cows at pasture will reduce the intake of herbage, and thus the time spent grazing. The supplement is to some extent self-defeating unless there is a shortage of available herbage. Weather does not have a major effect on grazing behaviour in temperate climates, apart from minor fluctuations due to driving rain and occasional hot days.

In the winter, eating occupies less time than grazing in the summer, and is about 4 to 7 hours per 24 hours. Cows self-feeding on low-D-value silage may have an average eating time of 4·4 hours per 24 hours, whereas cows with access to silage with a D-value of 70 in a trough may spend 6·3 hours eating. Concentrates are usually eaten rapidly, and an average intake of 0·4 to 0·5 kg per minute is normal for cubes of high acceptability and a diameter of 5 to 6 mm. Meals of a coarse texture are consumed at a slower rate than cubes (Fig. 17.2), especially when the amount on offer is increased. Meal mixed with water to form a soup-like mixture can be eaten four to five times more quickly than dry concentrates, but the practice has not been generally adopted.

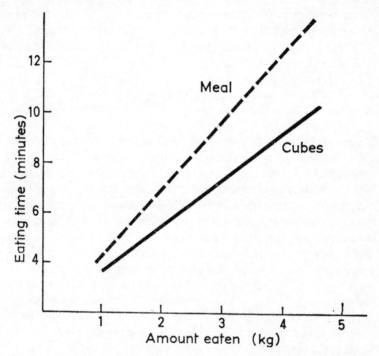

Fig. 17.2. Relationship between amount of concentrates and time of eating
(Adapted from C. G. Jones, *et al.*, in *Animal Production*, 8, 1966, p. 489)

Ruminating and Lying Down

Cows at pasture will ruminate for 5 to 9 hours per 24 hours, and about 80 per cent of this time occurs when the cows are lying down (Fig. 17.1). An average period of rumination will last about 30 minutes. On a typical winter ration of silage and concentrates, the ruminating time is about 8 hours per 24 hours. Cows lie down for 9 to 12 hours per 24 hours. The main period of lying down is in the night, and studies of cows in cubicles indicate that it is wise to have one cubicle per cow for herds of up to 40 cows, but that larger herds may have 105 cows per 100 cubicles. Cows in cubicles of the correct size are exceedingly tranquil when resting, although they will stand for a few minutes and then lie down in the same cubicle. When tied in a cowshed, the animals are more restless and regularly change from standing to lying. As a result, only 50 per cent of cows in a byre may be lying down at any one time whereas in cubicles, over 90 per cent of the animals may be lying down, especially in the early hours of the morning.

Walking and Other Activities

Cows on good-quality pasture will walk about 1,600 to 3,000 m per 24 hours, excluding the distance from the place of milking to the field, which is covered four times per day. If the grazing area available to the animals is increased, there is a tendency for the distance walked to increase. In areas of equal size, cows walk further as the yield of pasture decreases. It is thus advisable to offer the cows a sward with an adequate yield of herbage, about 2,000 kg dry matter per ha, and preferably in fields of restricted size. The distance walked while grazing is about equally divided between the periods between milkings, but with 80 per cent of the distance being covered in the hours of daylight. Since walking requires an expenditure of energy, there are sound reasons for ensuring that grazing fields are situated as close to the cowshed as possible and non-productive walking time is kept to a minimum.

At grass, dairy cows in milk will urinate and defecate about 10 and 12 times per 24 hours respectively. About 55 per cent of these excretions are made between the evening and morning milking, i.e. on the pasture grazed at night, compared with 35 per cent deposited on the pasture grazed during the day. This suggests a transference of fertility from pastures grazed during the day to pastures grazed at night. Urine and faeces deposited on the roadways, and in the milking area, comprise 10 per cent of the total excreted in the 24-hour period. On indoor rations in the winter, milking cows will urinate about 8 times, and defecate about 16 times per 24 hours. Cows which are not milking will have 15 to 30 per cent fewer excretions per 24 hours than cows in milk, owing to the lower intake of food and water.

Cows normally drink on 3 to 5 separate occasions per 24 hours, and the importance of an adequate water supply is discussed fully in Chapters Four and Eight. It is emphasized that a shortage of drinking water either at pasture in summer or in the buildings in winter can seriously affect the behaviour of a herd of cows, with a lowering of both feed intake and milk yield.

Social Rank

Within a herd of cows there are animals of high rank who dominate cows of a low rank. Animals in their first lactation tend to be the weakest, and older cows the strongest, in a system of ranking. Social rank is not of great importance if the animals are housed in a cowshed, but it can have a marked effect under conditions of loose housing. If silage is offered *ad libitum* in loose-housing systems, cows with a high rank will have a higher intake than cows of a lower rank. Milk production in the different groups may not be different but liveweight gain will be higher with the cows of higher rank. There can thus be large variations in silage intake and total animal production due to rank, even under good conditions of loose housing. Adequate, but not excessive, space will diminish the effects of rank, and is important in loose-housing conditions. Every effort should be made in the planning of buildings to ensure that animals are not restricted by social rank and poor building design from consuming an adequate amount of food. If water troughs are inadequate or badly sited, a few 'bully' cows can deprive weaker cows of an adequate water intake. The problem can be corrected by resiting the watering places, or by providing more drinking space. Water troughs are preferable to water bowls in systems of loose housing. Social rank is of less importance when grazing than indoors, but problems of bullying and shortage of water can occur in small paddocks where there is not sufficient individual space for each cow.

Normal Health

In order to detect ill-health in cows it is imperative to be familiar with the normal signs of good health. A healthy cow will eat a wide range of foodstuffs, and any refusals of food are often an early indication of ill-health. If the food itself is unpalatable, this will be indicated by a refusal in a high proportion of the herd. Dirty troughs may cause refusals of feed. Concentrates are often the first food to be refused, and this should be noticed when feeding in either the parlour or the cowshed. A lack of water, and foreign bodies in the mouth will cause food refusals, but this symptom may indicate a wide range of problems. There may be a tendency for a sick cow to isolate herself from the remainder of the herd and not to join in the main activity of the other animals.

Dung

The dung from a normal healthy cow is reasonably soft, with a dry-matter content of about 12 to 15 per cent. The colour and consistency of dung varies with the diet, and this fact must be recognized when assessing the normal situation. On fresh spring grass, the dung is soft, with a dry-matter content below 10 per cent, but this is quite normal as there is so little undigested feed residue to be excreted. When silage of high D-value is offered, the dung is also soft. Sugar-beet pulp will darken the colour of dung, whereas cereal-based concentrates will cause a light-brown colour. When dung is hard and in sections, this suggests that food intake is low. Scouring, i.e. dung of an almost liquid consistency, is not normal and indicates an intestinal irritation. This may be due to errors in the ration or to some specific disease. Normal faeces should not be too loose, and should not contain blood or bubbles of gas. The colour and consistency of the dung which is produced under the local herd conditions should be observed so that any changes can be detected rapidly.

Other Indicators

A normal cow has a regular pattern of rumination, as indicated previously, and any changes in the function of the rumen and intestines will be shown by a change in behaviour. Abnormal cows may stand with their heads down, and may grind their teeth. Foreign bodies in the reticulum, a displaced abomasum or bloat can all cause distress and discomfort, and will change the normal pattern of eating and ruminating. Cows rarely eject boluses of food when cudding, so if this occurs with any regularity, the animal and the food must be examined.

The breath of a healthy cow may smell slightly differently as the food of the animal is varied, but in general the odour is pleasant. A sweet and somewhat sickly, penetrating smell is indicative of abnormal digestive conditions, usually of acetonaemia. A good sniff of the atmosphere in a cowshed or cubicle house in the early morning before the cows start eating is helpful in detecting any abnormal odours. Laboured breathing and grunting are not normal and may indicate either a respiratory or a digestive problem. A cow's nostrils are normally moist and free from mucus.

The normal cow produces urine of a light-yellow colour, and thus any dark, blood-stained urine would indicate problems. Excessive feeding of kale can cause a dark urine.

The skin of a healthy, well fed cow is soft and pliable, with a bright appearance. If the skin is dull and lacks lustre there may be problems due to parasites, digestive disturbances or food intake. In advanced illness the skin and coat are harsh and staring. The condition of the skin must not be confused with a loss of liveweight, but on occasions the two may occur together. A loss of liveweight in early lactation is not abnormal, but a poor skin condition can indicate either a nutritional or a disease problem. Any sudden and dramatic loss of liveweight is a positive indication of some abnormality.

Milk Yield

The most sensitive indicator of any abnormality in feeding or health is the animal's milk yield. Although other symptoms may be overlooked, a fall in yield of milk is generally obvious, and the reasons for the decline should be investigated. Milk yield acts as a barometer in denoting normal and abnormal conditions, and should be considered the most valuable indicator. Changes in the amount of milk produced daily from the entire herd are generally a reflection of changes in herd numbers or the level of feeding, and should normally be easy to check and to explain. If all the cows in a herd decline in milk yield, the most probable explanation is a change in nutrient intake, either as a direct result of a feed change or indirectly because of some climatic effect. Marked decreases in milk yield can occur when cows are changed from a silage of high D-value to one of low D-value, or after a day of cold rain when grazing is restricted. Conversely, total milk yield will usually increase when cows are changed from winter feeding to spring grazing.

However, if the milk yield of an individual cow falls much more rapidly than that of other, similar animals, the reason should be investigated at once. Irregular milk yields can occur when the cow is in season, but all abnormal declines require a close examination of the udder and the milk for signs of mastitis, followed by a thorough scrutiny of the animal's normal functions. Lameness, digestive disorders, mastitis and many other abnormal conditions will all dramatically reduce milk yield in less than 24 hours.

At this stage it is useful to note the temperature of the cow by inserting a blunt-nosed thermometer into the rectum. Considerable variations can be found in the temperature of normal cows under different conditions, but the average range is 38 to 39 °C. Sick animals should be separated from the remainder of the herd and either put in a paddock near the farm buildings or housed in a well littered loose box, depending on the suspected condition. The diagnosis and treatment of disease is the job of the veterinary surgeon, who should be given a full account of the cow's history and symptoms.

Preventive Medicine

With co-operation between the farmer and the veterinary surgeon it is possible to reduce the incidence of, and even prevent, many diseases on the farm. This is both preferable to, and cheaper than, curing the trouble after it has arisen. For example, husk in calves can be prevented by dosing with vaccine before the animals are turned out to pasture, and the incidence of virus pneumonia is reduced if the buildings are correctly ventilated. The treatment of cows' feet by trimming and the use of a footbath is the best way of preventing lameness. Matters such as the fertilizer policy on the farm, and the milking routine, are all directly linked with animal health and require prior discussion if disease is to be kept at a low level. Accidents and diseases can occur at any time, but it is well worthwhile to take as many preventive measures as possible.

Mastitis

Mastitis is an inflammation of the udder associated with microbial infection and physiological change. The disease may result in either temporary or permanent damage to the milk-secreting tissue of the udder, and lead to a serious reduction in milk yield, a decline in milk composition and often a shorter life in the herd. Most clinical, and

95 per cent of the sub-clinical mastitis is caused by staphylococci and streptococci, although coliform and other pathogenic organisms can be implicated. Most cows calving for the first time are uninfected, but once an animal enters the herd there is a considerable and immediate risk of infection. This enters the udder via the streak canal of the teat. The degree of infection may vary from a sub-clinical form which may not be detected, to a severe clinical form which is recognized by clots in the milk, a reduced yield of milk and signs of heat and swelling in the udder. Mastitis is common, and at any one time 40 to 50 per cent of cows may be expected to have at least one infected quarter, and 1 to 2 per cent of all cows to show signs of clinical mastitis. The loss of milk due to mastitis is estimated to be at least 10 per cent of total production.

Cell Counts

Mastitis is regularly monitored by the Milk Marketing Boards, using a cell counting technique to measure the number of cells per ml of herd bulk milk. Inflammation in the udder causes an increase in the somatic cell count of the milk, and the number of cells is closely related to the amount of mastitis in the herd. An average cell count of 250,000 or less is satisfactory and should be the aim, but a count not exceeding 500,000 per ml is acceptable, although some herds with this count may have some clinical mastitis of short duration. Counts higher than 500,000 per ml, and where there is a progressive increase in the cell count, suggest that a mastitis problem exists, and that action should be taken. Counts over 1,000,000 indicate a severe mastitis problem.

Mastitis cannot be eliminated completely on a commercial farm, and eradication is therefore impracticable. However, it is possible to control the disease by a system of prevention which substantially reduces the number of new infections. The major control measures are the maintenance of the milking machine to the correct standard (Chapter Ten), a good milking technique to reduce udder damage, and a strict system of hygiene.

Control Hygiene

Adequate preparation of the cow by washing and drying of the udder just before milking will remove dirt and bacteria, but equally important, it will also stimulate the let-down of the milk by the release of the hormone oxytocin into the bloodstream. Communal udder cloths for udder washing must be avoided. A hose with hot running water and added iodophor or hypochlorite is preferable. Drying with a separate paper towel for each cow is also necessary. The foremilk should be drawn and examined for abnormalities such as clots, so that milk from any infected quarter can be isolated and the quarter treated. Where possible, first-calf heifers should be milked first, then cows free from mastitis, and finally animals with mastitis.

After milking, the teats must be immediately and carefully dipped in a specially formulated solution of chlorhexidine, hypochlorite, or iodophor. This dip will destroy organisms which may have been transferred to the teats during milking. Teat dips usually contain emollients such as lanolin to prevent sores and chapping. The instructions of the manufacturer should be followed carefully, as some products require dilution. The teats farthest from the milker should always be dipped first, and hand-held sprays are not recommended. The teat dip and udder wash should preferably be based on the same disinfectant, and should be changed periodically to lessen the risk of some bacteria becoming resistant to one product.

All animals should be routinely treated at the time of drying-off with an antibiotic which has been specifically formulated for this purpose. This simple routine measure does not contaminate saleable milk with antibiotic, and it will remove 60 to 70 per cent of existing infections and will prevent most new infections that are common in the early part of the dry period. In addition, any cows with clinical mastitis at any stage of the lactation should be treated with an antibiotic. The specific antibiotic for use during lactation and for dry-cow therapy should be discussed with the veterinary surgeon who knows the herd.

Old cows with clinical mastitis can be a potential source of infection in the herd, and should be culled. Such animals are probably high yielders, but the risk to the remainder of the herd is often greater than the value of the extra milk which they produce.

The benefits from a hygiene system are not immediate, but will increase slowly for at least 4 years (Fig. 17.3). Over this period the financial return should be more than double the costs of the system, and in subsequent years the margin will be even greater.

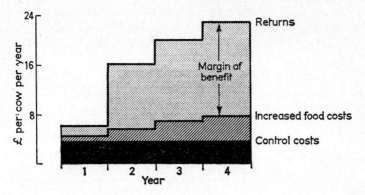

Fig. 17.3. Margin between benefits and costs per year in a 4-year mastitis control system
(From *Machine Milking*)

Summer Mastitis

This condition is due mainly to *Corynebacterium pyogenes*, and can affect milking cows, dry cows and heifers. It is thought to be spread by flies; the acute form results in the secretion of copious amounts of foul-smelling pus. Affected quarters rarely recover, and deaths of infected animals may occur. Control measures include spraying to reduce the fly nuisance, painting Stockholm tar on the teats of dry cows and regular inspection of the stock, but sporadic outbreaks will occur, especially in hot, dry summers. When a quarter has been treated and the cow has recovered it is often wise to cull the animal.

Milk Fever

Milk fever, or hypocalcaemia, is due to a sudden and temporary decline in the calcium content of the blood, and usually occurs within 48 hours after calving. The main symptoms are an unsteady gait and paddling of the hind feet; eventually the cow will lie down and not rise. If the cow is not treated, coma and death will often result. In the past, an excess of calcium was fed as a preventive measure before calving, but the recent understanding of the factors controlling calcium balance in the blood now indicates that a reduction, and not an increase, in the calcium intake before calving is a means of controlling this metabolic disease. If calcium-rich concentrates are given to cows on autumn pasture with a high calcium content, there can be an excess of calcium and hence a high incidence of milk fever. Prevention requires a correct balance of minerals in the diet.

A ration for the 4- to 5-week period before calving should supply about 50 g calcium and 30 g phosphorus per day. These intakes can be supplied from a normal ration of forages, and the major practical problem is how to increase the intake of energy and protein without increasing and unbalancing the mineral supply. The concentrates should be as low as possible in calcium and phosphorus, and normal dairy concentrates should only be given 2 to 3 days before calving. A further problem with this planned mineral control during the dry period is that reserves of minerals may not be replenished following a high lactation yield. Adequate concentrate feeding at the end of lactation before milking ceases is thus important. During the 4- to 5-week period of minimal calcium intake, the cows increase their efficiency of calcium absorption. Thus, at calving, when calcium levels in the blood fall, calcium mobilization from the bone and absorption from the intestine is more rapid and will help prevent the onset of milk fever.

To operate this control scheme it is essential to know the exact dates of calving. If the dates are not known it is advisable to increase the level of calcium feeding when calving is imminent and colostrum is available in the udder. It is better to delay the time of this increase until close to calving rather than feed a high-calcium ration for too many days before calving. The correct level of magnesium must also be in the ration to ensure that the calcium control system works. On some farms the calcium level in the forages may be excessively high, and such forages should be partly replaced by low-calcium cereals. There is increasing evidence that the incidence of milk fever can be reduced substantially by housing cows in the autumn and offering grass silage indoors rather than grazing outside.

Correct dietary balance is the first line of defence against milk fever, but a second line is the injection of high doses of vitamin D or an analogue a few days before calving. These injections must only be given under veterinary supervision, because of possible inherent dangers.

Treatment of milk fever is by a subcutaneous or intravenous injection of a sterile calcium borogluconate solution, plus phosphate and magnesium if required.

Grass Staggers

Staggers, or hypomagnesaemia, is characterized by a subnormal level of magnesium in the blood, and is generally associated with grazing. The symptoms are a reduced milk yield, nervousness and muscular tremors. In the acute form, the cow will stagger around, froth at the mouth, collapse and die. The primary cause of hypomagnesaemia in the grazing cow is a nutritional deficiency of magnesium, which is affected by the intake of dry matter, the magnesium content of the dry matter, and the availability of the magnesium. All these three factors can be low under certain combinations of grassland management and environmental conditions. Cattle must have a regular supply of magnesium, as storage in the body is low.

To prevent hypomagnesaemia it is necessary to increase the intake of magnesium, which can be done by careful grassland management (Chapter Four) and the feeding of extra magnesium. To increase the magnesium content of herbage it is advisable to apply magnesium limestone rather than ordinary limestone, and to increase the clover content of the sward. Applications of potash fertilizer and slurry rich in urine will lower the magnesium content of herbage and should not be given to pastures in the spring before grazing starts. Small applications of potash should be given in mid-summer. Pastures may also be dusted with calcined magnesite immediately before grazing; when strip-grazing, this task can be integrated with the moving of the electric fence.

The daily feeding of 60 g of calcined magnesite, i.e. magnesium oxide, per cow will prevent hypomagnesaemia, but acceptability of this dusty material can be a problem and it is preferable to offer the magnesite in a compound cube. 'Bullets' of magnesium alloy, administered via the oesophagus, will lodge in the reticulum and are also effective. In practice, however, some bullets are regurgitated at pasture, leaving cows at risk. Mineral licks containing magnesium are of little use.

Stress factors such as cold, wet weather, lack of shelter, rough handling, lack of food and being in season can all contribute to hypomagnesaemia.

Acetonaemia

Acetonaemia or ketosis is characterized by a reduced appetite, lowered milk yields and a sweet, pungent odour of pear drops in the breath and the milk. The condition is non-fatal, but 200 to 300 kg milk per lactation can be lost, and there will be a marked reduction in the body condition of the cow. This disease is associated with a shortage of glucose, i.e. hypoglycaemia, within the animal, due to an interference with the metabolism of fat and carbohydrate, and may be in either a sub-clinical or clinical form. The symptom of this is an accumulation of ketones in the blood. Sub-clinical ketosis is not easy to detect unless tests are made on the milk and the blood, but the condition is fairly widespread in the winter in recently calved cows. Milk yields are lowered, but there are no other clearly marked symptoms. The clinical condition is much more obvious: the appetite is reduced, the dung of the affected animal is hard, the yield of milk falls rapidly and the skin is tight and dry.

The disease will slowly disappear as milk yield falls, but this natural course will result in a large loss of milk. Prevention is based on supplying the animal with a balanced and acceptable ration containing sufficient digestible energy. A ration of low D-value silage and few concentrates can result in a 'starvation' ketosis, whereas a higher-quality ration containing adequate amounts of dry matter, energy, and protein will prevent the onset of the disease. Ketosis rarely occurs on top-quality grazing when the intake of nutrients is high. In the winter, ketosis is most likely to occur when the cow is newly calved and is drawing heavily on her bodily reserves to produce milk. At this crucial period, it is vital to ensure that the cow is well fed and has a high ME concentration in the total ration. Prevention of ketosis is preferable to treatment which can include injections of glucogenic steroids.

Ketosis as a herd problem is clearly associated with nutrition, but certain animals appear to be more susceptible than others. Careful husbandry and smooth changes in feeding are the key to lowering the incidence of this non-fatal but economically serious disorder.

Bloat

Bloat or rumenal tympany is a distension of the rumen by gas, and can rapidly be fatal. In the normal process of digestion, large volumes of carbon dioxide and methane are produced in the rumen of the cow, and after eating, 30 l gas per hour can be released by belching. Under some conditions the cows do not belch, and hence gas pressure will build up: and the resultant bloat is visible as a tight bulge on the left side of the animal. Bloat can be caused by an obstruction in the oesophagus, but is more commonly due to the diet of the cow containing less than the optimum of 16 to 20 per cent crude fibre.

Cows on lush, clovery pasture can rapidly develop bloat, and this condition is a serious problem in New Zealand, where the swards contain a high proportion of clover and are grazed tightly. Bloat is more likely to occur on legumes than on pure grass swards. Kale and turnips, especially on frosty mornings, can also produce bloat.

There are two types of bloat. In the free-gas type, gas accumulates above the food in the rumen, and it can be released by inserting either a sharp trocar and cannula or a knife into the highest point of the rumen. In frothy bloat, the gas is mixed with the digesta in the rumen and cannot be released by a puncture in the rumen. Treatment is by drenching, either with anti-foaming silicone compounds, which reduce the surface tension in the ingesta, or with a mixture of 750 ml raw linseed oil plus 30 ml turpentine. Treatment should be rapid if the animal is in a distressed condition, but great care must be taken when drenching a sick animal.

To prevent bloat, dairy cattle should not be allowed access to large areas of rapidly grown herbage, especially if legumes are present in large amounts. In this situation the animals will tend to eat the most succulent and least fibrous parts of the plants, and it is thus a sound policy to restrict the grazing area by an electric fence so that the cows will eat the whole plant including the fibrous stem. Hay and straw can be offered to cows at pasture to increase the intake of fibre, but consumption can be highly variable, with some cows not eating any. In New Zealand, peanut oil is sprayed on the pasture, and anti-bloat agents are metered into the water troughs in the field. Frosted food, in particular kale, should not be offered to cows.

A rapid intake of a large amount of concentrates and grain can also cause a type of bloat, and animals should be introduced slowly to this type of ration. Cows may accidentally gain access to a store of concentrates or grain, and problems of bloat similar to the 'feed-lot' bloat in the USA can occur. Bloat is not commonly a major problem, but nonetheless, because of its rapid onset and often fatal consequences, it is necessary to be constantly alert to observe the first symptoms in individual animals.

Lameness

Highly productive cows must have sound feet and legs. Breeding and selection can eliminate many unsound animals from the herd, but day-to-day management is far more important. Cows' feet should never be allowed to overgrow, and paring and filing of feet should be done as required. A crush is invaluable for this routine operation, but at times the animal must be cast by a suitable roping technique. A sharp pair of blacksmith's pincers and a sharp paring knife with a rounded end are required.

Lameness can arise because of uneven stony roadways, kale stumps in frosty weather and muddy fields. It is thus wise to improve cow roads and to avoid any areas which may aggravate foot problems. Walking on slightly roughened concrete will assist in keeping cows' feet in order, whereas housing in strawyards can accentuate the problems of overgrown and soft feet.

On occasions, an excess of protein in the diet will cause heat and swelling just above the hoof, but this fairly rare condition can be remedied by altering the ration.

Foul-in-the-foot, caused by *Fusiformis necrophorus*, occurs in the space between the digits and occasionally in the bulb of the heel, and may be the result of damage by stones and mud. If it is untreated, there can be serious lameness, with deep abscesses of the feet, secondary infections and the formation of pus. All offending bodies such as stones should be removed, the pus released and the wound dressed with an antiseptic. An injection of a sulphonamide, usually sulphadimidine, is also important to prevent abscesses in the deeper structures of the digit. The complete amputation of one of the two digits may be necessary to save a valuable breeding cow, but is of doubtful worth in a commercial situation.

Damage to the sole of the foot can lead to foot abscesses and infections of the bones and joints, as with foul-in-the-foot. Damage arises most commonly because of overgrown feet, which cause pressures and hence cracks which allow infections to enter.

In a cowshed, the problem of sore legs and hocks can be alleviated by the use of rubber mats. If cows are loose-housed, a shallow foot-bath, through which the cows are driven, is invaluable. The bath may be used once per week, with either a 10 per cent solution of copper sulphate or a 5 per cent solution of formalin.

Correct foot care will prolong the useful life of many cows, and will prevent the large losses of milk which result from lameness.

Metabolic Profile Test

A metabolic profile test is a herd test designed to reveal any abnormalities in the blood chemistry of the herd. Briefly, groups of cows at different stages of lactation are blood-sampled, and a range of blood components which are involved in the major metabolic pathways are determined rapidly. These components include blood glucose as a representative of energy metabolism, urea and albumin to indicate protein metabolism, and a range of mineral elements. The values of each component are then compared with 'normal' values. Interpretation of the test requires a full knowledge of the farming system, the condition of the cows, milk yield and stage of lactation, since some non-nutritional factors and seasonal variations can complicate an assessment of the results. The results of the tests need skilled interpretation by considering each component separately, and then the relationship of two or even three components to each other.

For example, abnormalities which can be revealed by the test are hypomagnesaemia, which must be rectified by feeding more magnesium, and hypoglycaemia, a lack of blood glucose, which indicates insufficient energy in the diet. This factor is linked with disappointing milk yields, and, as mentioned earlier, with ketosis. It is emphasized, however, that even the most sophisticated diagnostic test is only a guide and an aid, and therefore the results from a metabolic profile test must be used in conjunction with many other observations and records made on the specific dairy farm under consideration.

Further Reading

Bramley, A. J., Dodd, F. H. and Griffin, T. K., *Mastitis control and herd management*, 1981, NIRD-HRI Technical Bulletin 4, National Institute for Research in Dairying, Shinfield, Reading

Payne, J. M., *Metabolic diseases in farm animals*, 1977, William Heinemann Medical Books Ltd, 23 Bedford Square, London

Payne, J. M., Hibbitt, K. G. and Sanson, B. F. (eds.), *Production disease in farm animals*, 1973, Baillière Tindall, Henrietta Street, London

Sainsbury, D., *Animal health*, 1983, Granada Publishing, St Albans, Herts

Waite, R., 1963, 'Grazing behaviour', p. 286, *Animal health, production and pasture*, 1963, Longmans, Green and Co Ltd, London

CHAPTER EIGHTEEN

Business Aspects

Measures of Efficiency—Stocking Rate—Milk Sold Per Cow—Concentrate Use—Margin Over Concentrates—Factors Affecting Profitability—Capital Investment—Herd Size—Large Herds—Small Farms—Seasonality of Calving—Summer Versus Winter Milk Production—Choice of Breed

---

Dairy farming, like any other branch of farming—in fact more so than most—involves the employment, and perhaps the borrowing, of large amounts of capital. Good husbandry and stockmanship, while as important as they have ever been, are no longer enough in themselves to ensure prosperity; they have to be supplemented, in today's conditions, with at least an elementary understanding of business principles. A striking illustration of this is the approach to borrowing, which can have as dramatic effect on the fortunes of a farm as the stockmanship. Many a good dairy farmer has failed through imprudent borrowing—at the wrong time or for the wrong purposes—and many another has remained on a treadmill of physical work through reluctance to borrow when it was not only justifiable but clearly necessary.

There is probably no type of farming in which there is such a wide range of physical and financial performance as there is in dairying. To select just one financial criterion, gross margin per ha, the costings in Table 18.1 show a remarkable gulf between the top 25 per cent of herds and the bottom 25 per cent. It is, incidentally, the existence of this gap which gives encouragement to those who are either in the top 25 per cent or believe themselves capable of entering it.

Table 18.1. Physical and financial results from 1,130 dairy herds, 1981–2

	Average of all herds	Top 25% of herds	Bottom 25% of herds
Physical results			
Herd size	108	119	94
Milk yield (kg per cow)	5,272	5,748	4,726
Concentrates (t per cow)	1·76	1·82	1·71
Concentrates (kg per kg milk)	0·33	0·32	0·36
Stocking rate (cows per ha)	2·00	2·47	1·57
Fertilizer nitrogen (kg per ha)	246	299	190
Financial results			
Gross output* (£ per cow)	733	808	645
Variable costs† (£ per cow)	343	343	343
Gross margin (£ per cow)	390	465	302
Gross margin (£ per ha)	780	1,149	474
Margin over concentrates (£ per cow)	478	543	404

*Milk sales + calf sales − herd depreciation
†Concentrates + purchased bulk feed + forage + sundries
(Based on *An analysis of FMS costed farms*, 1981—82)

Measures of Efficiency

The only really accurate way of assessing the profitability of a farm is to ascertain the net profit after all outgoings other than income tax have been deducted, and relate this profit to both the size of the farm and the capital involved. Although this method may be of great interest to the individual farmer involved, fixed costs, and in particular finance charges, vary so much that to compare one farm with another on this basis is so laborious and complex as to be of doubtful value.

It has thus become customary to use a series of much simpler criteria as yardsticks for comparing the performance of one farm with that of others. These are either simply physical (e.g. litres of milk sold per cow) or financial but related only to the variable costs, i.e. those costs which vary according to the size of the enterprise. In other words, these financial criteria are concerned with gross margins rather than net profits.

Stocking Rate

Stocking rate is a measure of the area of land required to keep a cow, and may be expressed either as cows per ha or as ha per cow. For this purpose, an adult cow is taken as 1 'cow equivalent' and first- and second-year heifers as 0·4 and 0·6 cow equivalent respectively. Stocking rate is calculated by totalling the number of cow equivalents on the farm and dividing the sum into the number of ha, or *vice versa*.

There is no optimum stocking rate for all farms, although many efficient dairy farms have stabilized at about two cows per ha (or 0·5 ha per cow). In general, as the stocking rate intensifies, gross margin per ha increases but gross margin per cow decreases slightly and fixed costs, i.e. labour, machinery, and capital investment in livestock and machinery, increase.

A factor limiting stocking rate is often the desire to produce most, if not all, of the winter feed for the cows from the farm. On small farms, under about 35 ha, it may well pay to 'buy hectares' in the form of hay and other foods and accept a reduced gross margin per cow in order to maximize gross margin per ha. This is a good illustration of the importance of identifying the main limiting resource on a farm—on a small farm it is usually hectares—and ensuring that the maximum use is made of that resource. On larger farms, a very high stocking rate would be less easy to justify, since a more profitable use could perhaps not be found for the land released from the growing of fodder. On larger farms, a satisfactory return on the working capital employed in an enterprise is often of more importance than the maximum gross margin per ha.

Milk Sold Per Cow

This measure must not be confused with the average yield per cow as shown by the milk recording schemes, which is the average of all lactations of 305 days or less (excluding those under 200 days) completed during the year, regardless of the interval between calvings. 'Milk sold per cow' is calculated by dividing the total number of litres of milk produced from the herd during the year by the average number of cows and calved heifers in the herd during the same year. The latter figure is found by noting the number of animals in the herd on the first day of each month and averaging the 12 numbers. Milk sold per cow has the advantage that it takes into account the interval between calvings; if this interval increases, the figure of milk sold per cow decreases.

Milk sold per cow, like any other criterion of efficiency, means little by itself. Profitability depends not only on the number of litres sold, but on how economically they were produced. There is, however, a strong positive correlation between milk yield and gross margin per cow (Table 18.1), although the degree of this correlation will depend on the current ratio of milk price to concentrate price.

Concentrate Use

Concentrate use is expressed as kg per 1 milk or as t per cow. This measure is of great interest when considered in conjunction with the milk sold, but again means little in isolation. It is possible, for example, to feed a cow no concentrates at all, but only if a very low milk yield is acceptable.

While many costings tables suggest that concentrate use per litre does not rise as milk yield rises, this can be misleading, since there is a tendency for the herds with high yields also to be the more efficient. Concentrate use per 1 does tend to rise somewhat as lactation yield rises, if measured at a constant level of efficiency, but not to a great degree. It is impossible to predict precisely how much extra concentrates would be required to produce an extra litre of milk, since it depends on many different factors, including the genetic potential of the cows, their condition and health, and their present level of yield.

On average, higher-yielding herds tend to use more concentrates and produce higher gross margins than lower-yielding herds, but high gross margins do not necessarily accompany high concentrate use.

Margin Over Concentrates

This measure is probably the quickest, easiest and most generally useful way of comparing the efficiency of one herd with another. It is calculated by deducting the total cost of concentrates from total milk sales for the year and dividing this figure by the average number of cows and calved heifers in the herd. This measure takes into account both milk sold and concentrate use, and a high margin over concentrates may be achieved with either a high or a low milk yield. Margin over concentrates, taken in conjunction with stocking rate, is vital in determining the profitability of a dairy farm. However, like any other criterion, margin over concentrates must be treated with caution when looked at in isolation. A low concentrate cost may be achieved at an unduly high cost in labour, capital or land. An extreme example of high labour use would be the making of tripod hay, but a more usual one would be the hand-hauling of sugar-beet tops. An extreme example of high capital investment to reduce concentrate costs would be a farm-scale grass drier; a more usual example would be the use of highly sophisticated methods of making, storing and feeding silage. High land costs can arise from a low stocking rate or an overgenerous supply of reseeded grassland and catch crops.

There are other ways in which a high margin over concentrates can be achieved but farm profit may not be improved, such as a high culling rate, and calving heifers at 3 rather than 2 years old. Both these practices may be right in the context of a particular farm, but both can give a distorted value to this criterion.

A rise or fall in the price of concentrates should never be taken in itself as a reason for reducing or increasing the use of concentrates. What matters particularly is the relationship between the price of milk and the price of concentrates, and there are various ways of calculating this ratio. One useful method is to divide the price received for 1 l of milk in mid-winter by the cost of 0·4 kg of a typical dairy concentrate at the same date. The ratios in recent years have been as follows: 1980, 2·18; 1981, 2·24; 1982, 2.31.

Factors Affecting Profitability

Table 18.1 shows the physical and financial results from 1,130 dairy herds, costed by the Milk Marketing Board of England and Wales in the year between April 1981 and March 1982. The most striking aspect of the table is the huge range of performance between the top 25 per cent and the bottom 25 per cent of the herds. For example, the gross margin per ha for the top 25 per cent was £1,149, while that for the bottom 25 per cent was £474. The difference of £675 gross margin per ha is staggering, and the gap in performance is widening rapidly. The bottom 25 per cent of herds in Table 18.1 will be larger and more progressive than the lowest-performing herds in the country as a whole, which suggests that the true range of efficiency must be even wider than the figures indicate.

It will be seen that the top 25 per cent of herds were farming more intensively in every way—larger herds, a higher stocking rate, higher use of concentrates per cow, more nitrogen per ha, and producing more milk per cow and per ha.

Gross margins, as shown in Table 18.1, take no account of the level of investment with which the various results are achieved—in other words, of the level of fixed or overhead costs. Gross margins are a thoroughly useful yardstick for a quick comparison of the financial performance of different farms, but they should always be viewed against the background of the fixed costs. A herd with a gross margin well down in the bottom 25 per cent may be making as much profit as one in the top 25 per cent if no labour is employed, no money is borrowed, and there is little machinery.

The effect of fixed costs on farm profit is illustrated in the typical budget for a 100-cow herd in Table 18.2. From the net profit the farmer has to meet income tax and his living expenses, and make provision for future capital expenditure.

The budget in Table 18.2 illustrates clearly how the gross margin is determined largely by two items—milk sales and concentrate purchases. While every effort will naturally be made to control all other items in the accounts as well, it can be plainly seen that all other considerations are dwarfed by the importance of achieving the maximum margin of milk sales over concentrate costs. Whether this result is achieved by a large margin per cow from a comparatively small number of cows or a smaller margin from a larger number will not affect the gross margin, though it may well affect the net profit, since a larger herd involves more expenditure on cows, housing and labour.

Among the variable costs, herd depreciation is a figure which fluctuates widely from year to year according to the state of the beef market and may at different times be a plus or a minus figure. The profit of a herd with an average level of milk yield and an average culling rate will be affected to the same extent by any of the following price changes: 0·25p per l milk, £50 per cull cow, or £15 per calf.

The next largest item in the variable costs after concentrates is fertilizer, but it is small in comparison—less than 20 per cent.

Fixed costs vary from farm to farm even more than variable costs, due largely to the varying degree of mechanization and even more to the amount of borrowed capital in the business. Fixed costs include

Table 18.2. Budget for a typical 100-cow herd, 1981–2

Income	£	£
Milk sales	73,500	
Calf sales	5,900	
Less herd depreciation	− 3,400	
		76,000
Outgoings		
Variable costs		
Concentrates	24,400	
Purchased bulk feed	1,400	
Forage (fertilizer, seeds etc)	6,400	
Sundries (vet, litter, dairy supplies, AI etc)	4,800	
		37,000
Gross margin		39,000
Fixed costs		
Wages (paid)	6,120	
Power and machinery (costs)	7,560	
(depreciation)	3,000	
Sundries	3,080	
Property charges (costs)	3,800	
(depreciation)	2,480	
Interest	5,400	
		31,440
Net profit		£7,560

Note: average milk yield 5,300 l and concentrate use 1·83 t per cow
(Based on an average of figures from *An analysis of FMS costed farms, 1981–82*)

all those costs which cannot be allocated to any particular enterprise
and which do not vary in direct proportion to the size of the enterprise. Property charges include rent, rates, property repairs and
depreciation on the buildings and fixed equipment. Sundries include
water, insurance, fees and subscriptions.

If a dairy farm is not making a satisfactory profit, a useful procedure is to look carefully at each item of income and consider how it
could be increased, and then look at each item of expenditure and
see how it could be reduced. There is usually little that can be done
to improve calf and cull sales, and though there may well be scope
for savings in the variable costs, these can have little more than a
marginal effect on any except the two major items—concentrates and

fertilizer. It is unlikely, though not impossible, that reductions in either of these two items will increase profit. The answer to improving the gross margin will almost certainly be less a matter of financial control than of improving physical performance, for example increasing the stocking rate or improving the margin over concentrates.

The most difficult problems are those associated with fixed costs. Once these have risen, it is exceedingly difficult to reduce them. If there has been excessive expenditure on machinery, selling it will do little to help. If too much labour is employed, it is difficult and often socially unacceptable to reduce it, at any rate in the short term. If the farm is mortgaged and bank borrowing is high, little can be done to reduce these liabilities unless profits are being made. Where fixed costs are too high, the only answer as a rule lies in increasing production.

Capital Investment

Determining whether or not a proposed capital investment in buildings or machinery is worthwhile is a matter of careful budgeting of the total cost, allowing for any grants receivable, and weighing the net cost against the benefits to be derived from the investment. It is important to recognize that in an industry with such a rapid rate of obsolescence as dairy farming, some proportion of most investments must be regarded as periodic replacement rather than new investment. Money has to be injected into every dairy farm from time to time, partly in order to maintain its ability to compete in terms of technical performance and partly in order to provide the type of working conditions which modern workers expect.

A sound rule for borrowing is to match the source of the borrowing to the type of investment; for example, to use a mortgage for the purchase of additional land but a bank loan for an expansion of the herd. A second rule is to keep the proportion of the capital in the business which is borrowed within reasonable limits. This proportion is known as the 'gearing', and the gearing is high when the proportion of borrowed money is high. High gearing accentuates the effects of fluctuating profitability on a business; a highly geared business will make more profit when things are going well but will be much more vulnerable when profits fall. A final useful rule for borrowing is that the soundest investments usually either make increased production possible or make the productive process more pleasant for the men; investments aimed at reducing costs should be looked at with more caution.

Inflation can vitally affect investment decisions. The cost of a new dairy unit may seem very high at the time of building, but if building costs are inflating at 25 per cent per annum, the cost may look quite reasonable in 2 years' time.

Whether or not the investment appears to be worthwhile in the long term, the vital matter in the short term is that sufficient cash flow should be coming into the business to cover the cost of the borrowed money. If an investment has been made in order to increase production, it is crucially important that the increase should take place as soon as possible after the investment has been made.

The effect of excessive capital investment on fixed costs has already been stressed. On many farms there is scope for a saving on machinery costs, either by sharing with a neighbour or by participation in a machinery syndicate. There are also many jobs on a farm which can be more economically handled by a contractor than by the purchase of a machine which may stand idle for most of the year.

Herd Size

The size of a dairy herd may be limited by any of a wide number of factors: farm size, capital available, balance within the farm structure on a mixed farm, the capacity of the existing buildings, the ability of the man in charge, and above all by the throughput of the milking parlour. Throughput is determined by the type and size of the parlour, by the ability of the staff, by the cleanliness of the cows, and by the milking routine. Working on the assumption that 2·25 hours is a long enough period for a normal milking time (Chapter Eleven), herd numbers often settle at about 2·25 times the hourly throughput of the parlour. With a 10/10 herringbone parlour, a herd of 135 cows is quite usual, whereas with an automated 16/16 cow numbers may rise to 160 or 180. Precise numbers will of course be determined by many other factors, but the dominant one is very often the capacity of the milking parlour.

Having decided on an optimum size of herd for a farm, it is vitally important to maintain the herd number as near as possible to this size. It is unlikely to be good policy to exceed the optimum size, but a reduction in herd size can only cost money. For example, if the optimum herd size is 100 cows and the average number of cows over the year is only 94, the gross margin from 6 cows is being lost for a small saving on the interest on the capital cost of six replacement heifers. Maintenance of herd numbers is made more difficult by a seasonal calving policy.

There is no evidence that smaller herds are more efficient than larger herds, either in physical or in financial terms. There is no reason why increasing the size of a herd need impair its efficiency, provided that the facilities are improved at the same time so that the men do not become overworked.

Large Herds

A well equipped modern dairy unit may have 130 or more cows milked by one man, and a herd of this size has clear advantages over smaller herds; however, there do not appear to be further economies of scale above this size. For this reason, very large herds seem unlikely to make any significant impact in Britain. The factors which have moved the poultry industry into fewer, bigger units, and are doing the same for pigs, apply much less to dairy cows. Even a herd of 1,000 cows, for example, would scarcely eat enough concentrates to justify its own feed mill, whereas the animals would require huge quantities of bulky feed, which is difficult and expensive to transport. A herd of this size would also void great quantities of dung and urine in a form which is particularly difficult to handle. These problems all tilt the balance of advantage away from the large centralized unit and towards a number of one-man-milking units close to the source of feed, which in Britain is usually grass and grass products. The smaller units have proved to be easier to manage, and provide better job satisfaction for the dairy staff.

The principal advantage of large herds is the opportunity to group the cows closely by either yield or calving date, and to match the feeding of each group to its precise requirements. Large herds can also justify the use of more elaborate and efficient feeding machinery.

Small Farms

On small farms, which might be defined as farms which are not large enough to support a full-sized, modern dairy unit, special conditions apply. As on any farm, the first principle in establishing a farm policy is to identify the limiting resource and see that the best possible use is made of it. On a small farm, this is usually the area of land available, and the first priority must be to obtain the maximum return per ha, consistent with a sensible level of capital investment. Thus the tendency will be to intensify the stocking rate to the point where some or even all of the winter feed has to be bought in, and to feed for the maximum margin over concentrates per cow as well. A high output is the key to success on small farms.

Seasonality of Calving

A herd of cows may calve either within a comparatively short period (known as seasonal calving) or all round the year (known as a spread calving pattern).

In a seasonally calving herd, there need be no calvings at all at times of the year which do not suit the farm, and there is a quiet time of year when most of the herd is dry and holidays can be taken. It is also possible to manage and feed the cows more easily as one large group, though this advantage is reduced if the herd is large enough to be split into two or more groups.

On the other hand, with a spread calving pattern the heavy strain of having a large number of animals all calving within a short period is avoided, and the workload is spread more evenly round the year. Fewer calving boxes and calf pens are needed and, since milk production is more evenly spread, a smaller bulk tank can be used. The cowman's interest is also maintained better when cows are calving all round the year, and the calving of heifers at 2 years old is also much easier to organize, since they can be served when they reach the correct weight irrespective of the time of year. With seasonal calving, it is difficult to rear heifers from the cows calving later in the season—either spring or autumn—which may include some of the best animals. This restriction reduces the genetic pool for selection. In high-yielding herds there is less danger of milk-quality problems where peak lactation is not concentrated into a short period (Chapter Nine).

The balance of advantage appears to favour a spread calving pattern, but in fact most progressive herds have a reasonably close pattern, with heifers calving in August and September and the cows calving between September and January. A compelling argument against a spread calving pattern is often the difficulty of keeping spring-calved cows milking well at grass in late summer.

Summer Versus Winter Milk Production

Herds with extreme seasonal patterns of production do not, of course, produce *all* summer or *all* winter milk. The proportion produced during either the summer or the winter six months rarely exceeds 60 per cent of their total milk production. The contrast is thus one of emphasis. The main argument for summer milk production is the lower concentrate costs which are incurred when the spring-calved cow produces most of her milk from grazed grass. This is, however, the only cost which is significantly reduced in a spring-calving herd and against it must be set the lower milk yields produced by cows calving in spring (Table 18.3) and the lower price paid for summer milk. The relative profitability of summer milk production from year to year will depend on the weighting of the seasonal price schedule and on the price of concentrates.

Table 18.3. Lactation yield per cow as affected by month of calving

Month of calving	% change	Lactation yield, calculated from 5,000-l cow calving in January
January	0	5,000
February	−1·7	4,915
March	−5·3	4,735
April	−6·7	4,665
May	−5·0	4,750
June	−2·9	4,855
July	−1·7	4,915
August	+2·1	5,105
September	+3·6	5,180
October	+5·5	5,275
November	+7·8	5,390
December	+4·5	5,225

(From P. N. Wilson, Edinburgh Dairy Conference, 1977, East of Scotland Agricultural College, Edinburgh)

To be successful, a summer milk producer must be able to grow abundant grass and have the skill to utilize it efficiently. He must also have all his cows calving in a short period in early spring—usually between January and March—which carries with it all the advantages and all the drawbacks of extreme seasonal calving. In general, it has not proved easy to obtain good results from summer production, although there are some very efficient practitioners of the system.

Choice of Breed

Choice of breed is essentially a personal matter, but clearly no such choice can be made without reference to the financial implications. In order to evaluate the relative profitability of the various breeds certain basic information is required, including comparative figures for the breed's milk yield and composition, food requirements (which will also reflect the potential stocking rate), the average liveweight, the cost of replacements and the value of the cull cows and calves. There has been no large-scale experimental work in this country on the relative profitability of the breeds, and any attempt to assess this profitability on theoretical grounds should be treated with caution.

The performance figures in Table 1.7 give simple comparative results from recorded herds for various dairy breeds. Although there is no general relationship between the bodyweight of a cow and her milk yield, the average milk yield (Table 1.7) of the four main breeds and the average liveweights (Table 2.3) are directly related. The yield of fat is, however, unrelated to bodyweight, and the fat output from the Jersey is only slightly lower than that from the Friesian. The maintenance requirement of the Jersey (Table 2.3) is substantially lower than that of the Friesian, although the requirements per kg of milk (Table 2.4) are higher in breeds such as the Jersey and Guernsey which have milk with a high content of fat.

The low maintenance requirement of the Jersey means that this breed is the most efficient producer of milk, fat and milk solids per unit of both bodyweight and land. In theory income from milk will be greater, particularly when a premium is paid for fat and total solids yet, in practice, costings consistently show a lower gross margin both per cow and per ha for the Channel Island breeds than for the larger breeds such as the Friesian. The general preference of farmers is well illustrated by the fact that in England and Wales

88·6 per cent of the cows are Friesians and only 2·0 per cent are Jerseys (Table 1.4). The reaons why the theoretical advantage of the Jersey does not appear to be generally translated into practice are not entirely clear, but the explanation may lie partly in the superior beef qualities of the Friesian, and partly in a general failure to take advantage of the potentially higher stocking rate of the Jersey.

There are, of course, many herds of all breeds whose results differ markedly from the average figures which appear in the costings, and an expert stockman with sound business sense will make a success with any breed of his choice.

Further Reading

Amies, S. J., *Block calving the dairy herd*, Report No. 26, 1981, Milk Marketing Board, Reading

An analysis of FMS costed farms 1981–82, Report No. 33, 1982, Milk Marketing Board, Reading

Nix, J. S. with Hill, G. P., *Farm management pocketbook*, 12th ed., 1981, Wye College (University of London), Ashford, Kent

The agricultural budgetting and costing book, No. 17, 1983, Agro Business Consultants Ltd, Atherstone, Warwickshire

APPENDIX ONE

Nutritive Value of the Main Foods for Cattle

Food	Dry matter content (%)*	Metabolizable energy (MJ per kg DM)	Digestible crude protein (% of DM)	Digestible organic matter in DM† (%)	Calcium (% of DM)	Phosphorus (% of DM)
Grass						
Grazing, high-quality	20·0	12·1	18·5	75	0·66	0·30
Ryegrass, post-flowering	25·0	8·4	7·2	55	0·45	0·25
Green legumes						
Red clover starting to flower	19·0	10·2	13·2	65	1·76	0·29
Lucerne, early flower	24·0	8·2	13·0	54	2·10	0·40
Silage						
Grass, high-digestibility	25·0	10·2	11·6	67	0·75	0·35
Grass, low-digestibility	25·0	7·6	9·8	52	0·70	0·30
Red clover	22·0	8·8	13·5	56	1·54	0·21
Maize	21·0	10·8	7·0	65	0·37	0·32
Hay						
Grass, high-quality	85·0	9·0	5·8	61	0·40	0·25
Grass, low-quality	85·0	7·5	4·5	51	0·33	0·21
Lucerne, half-flowering	85·0	8·2	16·6	55	1·93	0·26
Dried grass and legumes						
Grass, leafy	90·0	10·6	13·6	68	0·75	0·30
Lucerne, early flower	90·0	8·7	12·8	57	2·00	0·26
Straw						
Barley, spring	86·0	7·3	0·9	49	0·34	0·09
Oat, spring	86·0	6·7	1·1	46	0·34	0·10

Roots					
Turnip	9·0	11·2	7·3	72	0·48	0·34
Swede	12·0	12·8	9·1	82	0·48	0·23
Sugar beet	23·0	13·7	3·5	87	0·16	0·19
Green crops					
Kale, marrow-stem	14·0	11·0	11·8	70	2·14	0·33
Cabbage, drumhead	11·0	10·4	10·0	66	1·36	0·27
Grain					
Barley	86·0	12·9	8·2	86	0·05	0·38
Maize	86·0	14·2	7·8	87	0·02	0·27
Oats	86·0	12·0	8·4	68	0·09	0·37
Oil cake and meals					
Coconut meal	90·0	12·7	17·4	74	0·22	0·66
Cotton cake (decorticated)	90·0	12·3	39·3	70	0·32	1·47
Groundnut meal (decorticated, extracted)	90·0	11·7	49·1	75	0·16	0·63
Palm kernel meal (extracted)	90·0	12·2	20·4	78	0·23	0·56
Soya bean meal (extracted)	90·0	12·3	45·3	79	0·23	1·02
Other foods					
Fish meal, white	90·0	11·1	63·1	68	7·93	4·37
Brewers grains, barley, fresh	22·0	10·0	14·9	59	0·50	0·59
Brewers grains, dried	90·0	10·3	14·5	60	0·32	0·78
Maize, flaked	90·0	15·0	10·6	92	—	0·29
Maize, gluten meal	90·0	14·2	33·9	85	0·04	0·14
Sugar-beet pulp, molassed	90·0	12·2	6·1	79	0·63	0·07
Beans, field	86·0	12·8	23·0	81	0·18	0·66

* To convert to g per kg, multiply by 10
† D-value

Content of Calcium, Phosphorus, Magnesium and Sodium in the Main Mineral Supplements

Mineral supplement	(%)			
	Ca	P	Mg	Na
Steamed bone	38·5	13·5	0·35	0·47
Dicalcium phosphate	23·6	17·9	—	—
Monosodium phosphate	—	17·0	—	25·7
Ground limestone	38·0	—	—	—
Common salt	—	—	—	39·0
Calcined magnesite	—	—	52·0	—

Note: the values in this Appendix are based on data in the following five publications, which include more detailed compositions of a wider range of foods for animals.

Energy allowances and feeding systems for ruminants, Technical Bulletin No. 33, 1975, HMSO, London

Feedingstuffs evaluation unit, First Report, 1976, Rowett Research Institute, Aberdeen and DAFS, Edinburgh

McDonald, P., Edwards, R. A. and Greenhalgh, J. F. D., *Animal Nutrition*, 1981, 3rd ed., Longmans Ltd, London

Nutrient allowances and composition of feedingstuffs for ruminants, ADAS Advisory Paper No. 11, 2nd edition, 1976, Ministry of Agriculture, Fisheries and Food, London

APPENDIX TWO

Addresses of Organizations Concerned with Dairying

Research Institutes
National Institute for Research in Dairying, Shinfield, Reading
  RG2 9AT
Hannah Research Institute, Ayr KA6 5HL
Agricultural Research Institute of Northern Ireland, Hillsborough
  Co. Down, BT26 6DP

Experimental Husbandry Farms
Boxworth EH Farm, Boxworth, Cambridge CB3 8NN
Bridget's EH Farm, Martyr Worthy, Winchester, Hampshire
  SO21 1AP
Trawsgoed EH Farm, Trawsgoed, Aberystwyth, Dyfed SY23 4HT
Crichton Royal Farm, Bankend Road, Dumfries DG1 4SZ

Bureau
Commonwealth Bureau of Dairy Science and Technology, Lane
  End House, Shinfield, Reading RG2 9BB

Milk Marketing Boards
England and Wales, Thames Ditton, Surrey KT7 0EL
Scottish, Underwood Road, Paisley PA3 1TJ
Aberdeen and District, P.O. Box 117, Twin Spires, Bucksburn,
  Aberdeen AB9 8AH
North of Scotland, Claymore House, 29 Ardconnel Terrace, Inver-
  ness IV2 3AF
Northern Ireland, 456 Antrim Road, Belfast BT15 5GD

Societies and Associations
Royal Association of British Dairy Farmers, Robarts House,
  Rossmore Road, London NW1 6NP
British Society of Animal Production, Milk Marketing Board,
  Thames Ditton, Surrey KT7 0EL

APPENDIX THREE

Metric and Imperial Equivalents

Length	
1 mm	= 0.04 inch
1 m	= 3·28 feet
1 km	= 0·62 mile
Area	
1 m²	= 10·8 square feet
1 ha	= 2·47 acre
Volume	
1 l	= 0·22 gallon
Mass (*weight*)	
1 kg	= 2·20 pounds
1 t	= 0·98 ton
Weight/area	
1 kg/ha	= 0·89 pound/acre
1 t ha	= 0·40 ton/acre
Pressure (*vacuum*)	
100 kPa (1 bar)	= 29·5 inches mercury
Temperature	
° C	= ($^{\circ}$F – 32) × $\frac{5}{9}$

INDEX

abomasum, 33, 36
abortion, mycotic, 96
acetonaemia, *see* ketosis
acid
    acetic, 35, 37–8, 102
    butyric, 35, 37–8, 102, 105
    formic, 105
    lactic, 101
    nitric, 171
    oxalic, 122
    palmitic, 133
    propionic, 35, 37–8
    sulphamic, 171
    valeric, 35
albumin, 281
alimentary tract, 32–3
alveolar cells, 37, 162–3
amino acids, 32, 36, 38, 102, 152
ammonia, 35, 36, 108, 109, 133
anaemia, kale, 121
anaesthetic, 257
antibiotic
    in milk, 159
    treatment, 274
antibodies, 255
artificial insemination, 26, 28, 241, 247
ash, 31, 152–3

barley, 31, 46–7, 87, 109, 130, 132
barrier
    diagonal, 139
    electric, 141–2
    kerb, 139
    tombstone, 139, 141
    yoke, 139, 183
bedding, 193, 194, 202, 212
beef
    bull, 240, 254
    production, 27–8
beet
    fodder, 116, 119–20
    pulp, 130, 247
    sugar, 120

behaviour
    eating, 265–6
    excretions, 268
    grazing, 265, 266
    lying down, 267–8
    rank, 269
    ruminating, 267, 270
    walking, 268
biological oxygen demand (BOD), 221
biuret, 134
bloat, 87, 279–80
blocks, feeding, 134
blood
    calcium, 275–6
    components, 281
    flow, 37
    ketones, 278
    magnesium, 277
    supply to udder, 162–3
body-condition scoring, 44, 111–12, 232–4
bolus, 270
bonus scheme, 234–5
book, Movement of Animals, 227
borogluconate, calcium, 276
borrowing, 283, 290
box
    bull, 208–9
    calf, 207–8
    calving, 206
    isolation, 206–7
breeding systems
    cross breeding, 239
    inbreeding, 238
    line breeding, 239
    out crossing, 239
breeds
    beef, 25, 254
    choice, 295
    dairy, 24–6, 153, 295
brewers'
    grains, 130, 135
    yeast, 130